THE SUN

Kosmos

A series exploring our expanding knowledge
of the cosmos through science and technology
and investigating historical, contemporary
and future developments as well as providing
guidance for all those interested in astronomy.

Series Editor: Peter Morris

The Sun

Leon Golub and
Jay M. Pasachoff

REAKTION BOOKS
Published in association with
THE SCIENCE MUSEUM, LONDON

Published by Reaktion Books Ltd
Unit 32, Waterside
44–48 Wharf Road
London N1 7UX, UK
www.reaktionbooks.co.uk

In association with
Science Museum
Exhibition Road
London SW7 2DD, UK
www.sciencemuseum.org.uk

First published 2017
Copyright © Leon Golub and Jay M. Pasachoff 2017
Science Museum ® SCMG Enterprises Ltd
and logo design © SCMG Enterprises Ltd

Printed and bound in China

A catalogue record for this book is available from the British Library

ISBN 978 1 78023 757 2

CONTENTS

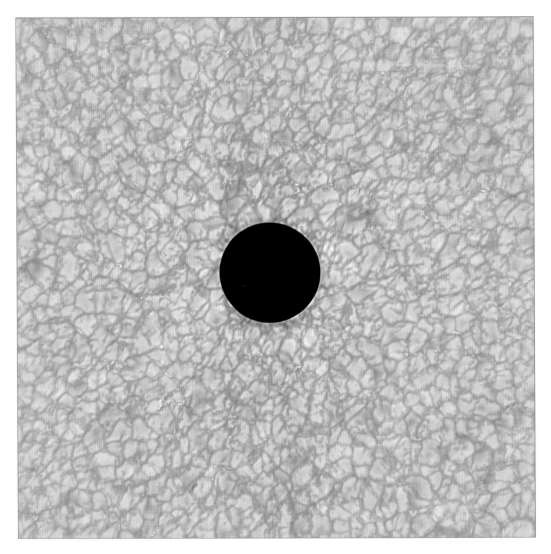

1 The silhouette of Mercury is suspended in front of the solar granulation in this image from the 9 May 2016 transit of Mercury across the face of the Sun. The granulation is structure roughly the size of the United Kingdom in the everyday surface of the Sun, which is known as the solar photosphere. The structure results from convection, which is similar to boiling. The image was taken with the 1.6-m (mirror's diameter) New Solar Telescope, with adaptive optics, at the Big Bear Solar Observatory in California by one of the authors (JMP) and colleagues. The apparent smaller structure adjacent to Mercury's disc is an artefact of the adaptive-optics process that makes the detail so sharp.

Preface

It is readily apparent that the Sun matters for life on Earth and that it has a major influence on our environment: from the warmth near the equator with the Sun overhead, to the temperate zones in which most of the world's population lives, to the icy lands of the midnight Sun, we see the effects of the extremes of the amount of sunlight on Earth. Every day we feel the effect of its coming and going – literally the difference between day and night.

But getting beyond that basic perception requires a scientific understanding. The Sun is not the same size and distance as the Moon, appearances notwithstanding. It does not revolve around the Earth, even though it looks as though it does. Figuring out what it is, what it's made of, why it glows so brightly, how old it is, how long it will last – all of these take much thought and experiment. The purpose of this book is to begin to get at some of that understanding, to explore what we know about the Sun and how we have come to know it.

There are many ways to approach this topic. Here we have chosen a slightly unusual way that we hope will complement the more traditional ones. We have carefully selected some of the most striking and important images of the Sun – some going back to the early seventeenth century, some reaching us daily, in near real time – using each as the focus of a discussion about the solar phenomenon in the picture. What is it that we are looking at? Why does it matter? What's important about this thing? What do we know about it, and how did we come to know it? What do we still not know? It may seem a little strange that a discussion of the interior of the Sun begins with a report of earthquakes in India, but this interconnectedness is actually a central feature of scientific research. There are no clean divisions between subjects in Nature – it is all one system.

The images we've selected range from the inside of the Sun (how is that possible?) to the surface, where the familiar sunspots are seen, out into the invisible (to our limited eyes) corona and solar wind, to the heliosphere, the volume of space controlled by the energy and mass flowing out from the Sun. The many meanings of the word 'picture' feature prominently, moving from a literal picture, to visualization, to the formation of mental schemata for understanding. We even include a spectrum – something rare in such discussions – and we show how the interpretation of this more generalized type of picture is crucial to our ability to see (with its multiple meanings) what is going on.

One of the major themes in the history of solar physics, and of astrophysics in general for that matter, has been the developing understanding of how widespread activity and dynamical variability are throughout all of known space. The universe is far more variable, dynamic and explosive than was realized just a few generations ago, with gamma-ray bursts, jets from compact objects, explosive events from the Sun and from other stars, and expanding magnetic shock waves from supernovae and other objects, to name just a few. In many cases magnetic fields are deeply implicated in causing the activity or in the effects of the activity. As one of our colleagues (there is some debate about exactly who said it first) said: 'Magnetic fields are to astrophysics what sex is to psychoanalysis.'

Observing the Sun can be dangerous to one's eyesight if not done with proper precautions and you should never look directly at the Sun through a telescope without special solar observing filters in place. In Appendix I we have included a discussion on 'Observing the Sun Safely' to guide people who would like to engage in solar observing. We also have an Appendix on observing the Sun at eclipse and, a bit less practical for most people, from space.

For those who are interested in exploring further, we provide some suggested reading at the end of the book. Many good popularizations of various subjects that we touch on in this book have been written recently, and we list them in the relevant suggestions for further reading. For those who like proof to back up the assertions we make, there is also a brief list of some of the more technical articles as well.

We are maintaining a website for updates and corrections to the book at:

http://web.williams.edu/Astronomy/sciencemuseumlondon.

It is also available through:

http://solarcorona.com.

We also provide a variety of links for websites relevant to the science of the Sun, observing eclipses and other matters that might be of interest to readers of this book. We note an article with such links: Jay M. Pasachoff, 'Resource Letter SP-1 on Solar Physics', *American Journal of Physics*, LXXVIII (September 2010), pp. 890–901.

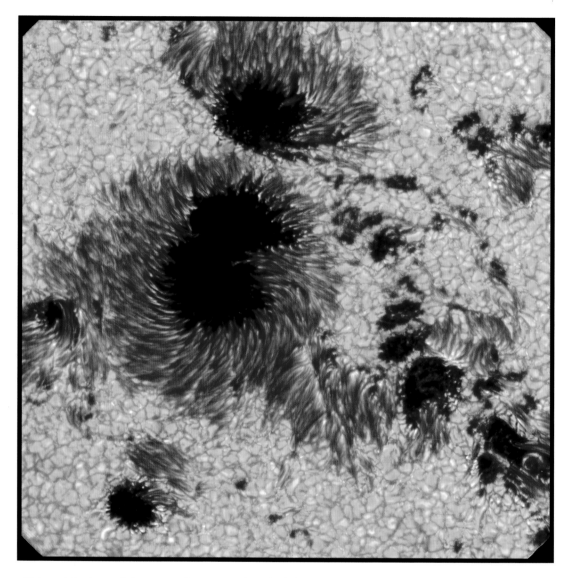

2 High-resolution image of a large group of sunspots. It shows the dark sunspot umbrae, each surrounded by a lighter ring of thin threads, more or less radial from the umbra and called the penumbra, all of this surrounded by the undisturbed photosphere, the granular pattern in the surrounding part of the image, representing the normal, widespread granulation seen all over the solar surface.

ONE

SUNSPOTS

Normally the Sun is too bright to look at directly without damaging the eyes, but when its brightness is diminished to a safe level by sunset haze or thin clouds, or if a special dark filter is used, we are able to see the round, yellowish circle of glowing light. At a quick glance it looks perfectly blank, but sometimes – for those with good vision – a small black dot or two can be seen marring the disc. With practice, groups of these dots can sometimes be seen. Repeated observation from one day to the next can show that the spots move across the disc, or grow or fade away from one day to the next. The obvious name to give these features is sunspots (illus. 2).[1]

Recent archaeological investigations, primarily of inscriptions on ancient bones, indicate that observations of the Sun were important in China dating back to at least the Shang dynasty (c. 1500–1050 BCE). Systematic recording of sunspots is known to have begun during the Han dynasty (from 206 BCE), although the reason for this early interest is not known and there are far fewer records than there ought to be if spots were the main item of interest. But many of the descriptions are clear and allow us to know that the Sun had spots back then, as it does now, so we know that sunspots are a normal, enduring feature of the visible Sun.

In the West, there don't seem to have been many sunspot records. Perhaps the ancient belief that the Sun was perfect and

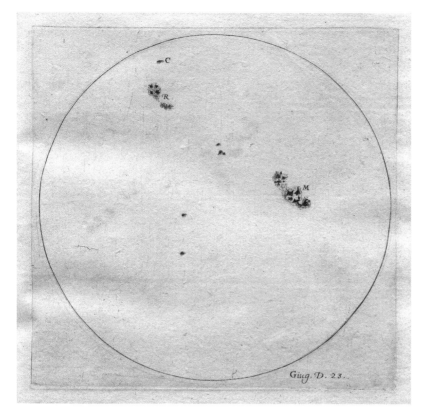

3 As we shall see, Galileo and several competitors were the first to see and record details of sunspots. This drawing is from Galileo's sunspot book of 1613, which shows a daily series of sunspots that reveals the Sun's rotation.

spotless discouraged any reporting of imperfections. But in the early 1600s Galileo changed things when he began using the recently invented telescope to observe celestial objects, including the Sun. He made careful, repeated observations (which may have contributed to his blindness) and drawings of sunspots as they grew, decayed and moved across the face of the Sun. From careful study of these sequences of drawings Galileo argued, from measurements of the path they follow across the disc, that sunspots are on the surface of the Sun and therefore are actual solar features. Fierce debates followed, arguing over whether the blemishes were indeed on the Sun, or passing in front of the Sun, or a type of cloud between us and the Sun.

Soon after these initial telescopic observations something strange happened: the spots vanished. For decades, from 1645 to 1715 and pretty much coinciding with the reign of Louis XIV ('le Roi Soleil' – the Sun King) in France, the Sun became nearly spotless. The absence of spots was well known at the time: the meticulous observer Johannes Hevelius noted in 1668 that 'for a good many years now, ten or more, I am certain that absolutely nothing of great significance (apart from some rather unimportant and small spots) has been observed.' The period of absent spots also coincides approximately with the deepest part of an extended spell of cold weather in Europe known as the Little Ice Age, leading to speculation that the absence of spots caused the Earth to cool somewhat. But the timing doesn't quite work out for that, since the cooling seems to have begun about 1550 and lasted until 1850, and indirect evidence from glaciations, snowfall and ice core records provides only slight support to the idea that the rest of the world cooled the way Europe did. So the strength of the connection between sunspots and climate is still uncertain and is problematic at best.

But then in the eighteenth century the spots returned and they have appeared regularly on the Sun ever since. They do come and go in cyclical fashion: there are years when the Sun is peppered with large numbers of spots and other years when there are hardly any. The change from the low 'solar minimum' state to the high 'maximum' state takes a few years, and overall the time from one maximum to the next is about eleven years, give or take a few from one cycle to another. This is a surprisingly short time for a star to show such obvious changes, when we consider that the Sun is otherwise steady, needing many millions of years for any significant changes to happen.

The sunspot cycle is one of the big mysteries about the Sun that we are struggling to understand. For now, we ask a more basic question: what are these spots? The answer to that question took a few hundred years to discover and turned out to be surprising.

Sir William Herschel was born Wilhelm Friedrich Herschel in Hannover in 1738, the son of an oboist in the Hannover Military Band (he later himself became a first-rate musician). In 1757, at the age of nineteen, he was sent to England, where his musical activities continued. It was not until the 1770s that he and his sister Caroline (whom he had convinced to join him in England as a singer at his concerts) actively turned to astronomical investigations. With Caroline's assistance, William discovered the planet Uranus in 1781, the first new planet found in over two thousand years. He named it *Georgium Sidus* – the Georgian planet – after King George III and both received lifetime pensions in return (£200 a year for William, £50 for Caroline), allowing them to devote themselves full-time to astronomy. Among his solar studies, William attempted to correlate the price of wheat in London with the variations in sunspot number, and he also discovered that the Sun emits vast amounts of radiation beyond the red end of the visible spectrum (now called 'infrared').[2] He proposed that sunspots are holes in the Sun through which one can see the dark interior, rather like the pupil of your eye. This followed a finding by the Scottish astronomer Alexander Wilson, Chair of Practical Astronomy at the University of Glasgow from 1760, who noticed that the dark spots appear to be slightly depressed below the visible surface of the Sun when they are viewed near the edge of the solar disc. It would be many years before the field of thermodynamics developed enough to show that Herschel's theory was not tenable, and the true nature of sunspots remained a mystery into the twentieth century.

The Modern Age of Sunspot Studies

A most significant step in understanding sunspots was taken by the energetic American scientist George Ellery Hale at the end of the nineteenth century and into the early twentieth century. Hale was born in Chicago in 1868 into an engineering family (his father

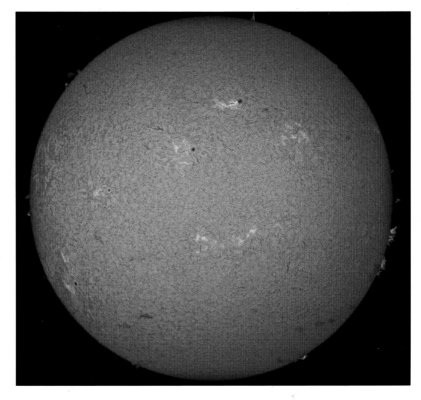

4 Spectroheliogram image of the Sun taken on New Year's Day, 1 January 2013, near the maximum of the solar activity cycle. This image uses a method similar to that pioneered by George Ellery Hale, a very narrow waveband centred on the strong emission line of hydrogen at 656.3 nm wavelength. Only a few small, isolated sunspots are visible, indicating that this maximum was weak compared to those of the past fifty years.

manufactured elevators, including the one in the Eiffel Tower) and studied at MIT. As a student there in 1889 he developed a new type of instrument for observing the Sun, the spectroheliograph, which is to this day one of the major tools for solar observation. Indeed, a version of this instrument was used to produce the sunspot image shown in illus. 4. Back in Chicago a few years later, he oversaw construction of the Yerkes Observatory, which included the largest refracting telescope (using lenses rather than mirrors) ever built. With support from Andrew Carnegie, Hale went on to build the Mount Wilson Solar Observatory in California, noting that 'The prime object of the Solar Observatory is to apply new instruments and methods of research in a study of the physical elements of the problem of stellar evolution.' To this end, he constructed laboratory facilities at his observatories to carry out experiments that helped to explain the astronomical observations, a field now known as laboratory astrophysics.

After developing Mount Wilson, Hale went on to plan the Palomar Observatory, which was for many years the largest astronomical telescope in the world. But Hale did more than build observatories. As a young professor at the University of Chicago in 1895 he founded the American Astronomical Society, the leading professional organization in the U.S. for astronomers, and the *Astrophysical Journal*, which became and remains one of the premier professional journals in the world for publishing research in astrophysics. In 1904 he organized an international scientific group that later became the International Astronomical Union, arguably the leading professional organization for astronomers in the world. In 1907 he joined the board of the Throop Institute in Pasadena and led the effort that transformed it into the California Institute of Technology. In 1916 he led the founding of the National Research Council, the working arm of the U.S. National Academy of Sciences that carries out most of its scientific studies.[3]

Hale had the idea of trying to determine whether or not sunspots are magnetized. Of course, the fact that he made the measurement at all implies that he thought the spots might be magnetic features. He thought the fields were probably generated by electric currents, in what he described in a 1908 issue of the *Astrophysical Journal* as 'vortices' around sunspots (illus. 5), although his description seems to indicate that he saw the spots as being similar to terrestrial cyclones: 'It seems evident, on mere inspection of these photographs, that sun-spots are centres of attraction, drawing towards them the hydrogen of the solar atmosphere.'

Whatever his motivation may have been, Hale decided to apply a recently discovered technique to measure magnetic fields merely by study of the light coming from a distant object. In 1896 the Dutch scientist Pieter Zeeman (Nobel Prize in Physics, 1902) had announced the discovery of a method for measuring magnetic fields in a hot gas by careful analysis of the light emitted by the gas. Zeeman showed that the energy levels in the emitting gas atoms would be slightly shifted due to the presence of a magnetic field, thus slightly altering the wavelength of the emitted light. He concluded his paper by suggesting that this method would be useful in astrophysics. Hale picked up on this idea by applying it to sunspots, and he struck gold.

5 Hale saw swirls and vortices around sunspots, convincing him that they were magnetic features. Referring to the spot shown here, he wrote: 'the clearly defined whirls point to the existence of cyclonic storms or vortices.' Note that the vertical stripes in the image are not present on the Sun, but are rather an artefact of the method used to create the spectroheliograms.

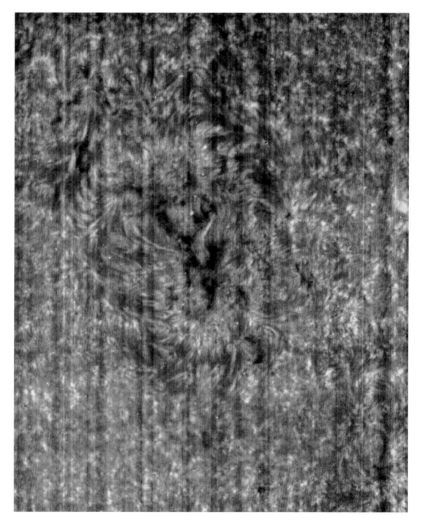

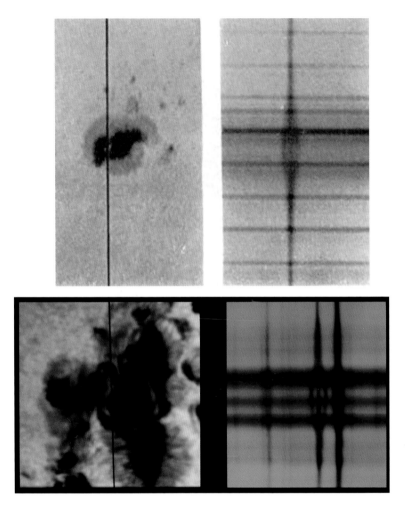

6 *Top*: The measurement that convinced Hale of the magnetic nature of sunspots is shown here, where the special pattern of wavelength splitting predicted by Zeeman is observed in the light coming from the sunspot. The vertical line on the left is the slit of the spectrograph, through which the light from the Sun passes on its way to the film. The image on the right shows the spectrum, the light that came through the slit having been spread out in wavelength. At the place where the slit crosses the sunspot, the image on the right shows the spectral line split into three components by the strong magnetic field. The horizontal lines are due to the construction of the filter used to make the measurements. *Bottom*: A modern version of the same measurement, from the McMath-Pierce Solar Facility on Kitt Peak, Arizona.

What did Hale see that he found to be such convincing evidence of a strong magnetic field? Illustration 6 shows an example, from Hale's 1919 paper, of the so-called Zeeman Effect in a sunspot. A telescope focuses an image of the Sun onto a shiny plate that has a slit cut into it, and the slit is positioned across the location of a sunspot. This is shown in the left half of the figure, where the dark vertical strip is the slit through which some of the incoming light can pass, including light from the darker sunspot located near the centre of the image. The light coming through the slit is then analysed by spreading out the different wavelengths in the light, using a specially chosen wavelength that is sensitive to the presence of a magnetic field. The result is shown in the right half of the figure, where the wavelength-spreading is in the left–right direction. It is easy to see just by looking at it that the light is doing something strange in the area of the sunspot. Because the magnetic field causes shifted energy levels in the atoms of the solar gas, the wavelengths of the light emitted by the atoms are split into several components, exactly as described by Zeeman in his laboratory experiments. This splitting of the wavelengths, as well as other detailed properties of the light (the polarization, for instance, at different viewing angles), shows that the light coming from the spot is being produced in a place that has a strong magnetic field, several thousand times stronger than the average magnetic field of the Earth, for instance. Zeeman himself wrote a commentary on Hale's article: 'Professor Hale has given what appears to be decisive evidence that sunspots have strong magnetic fields, the direction of these fields being mainly perpendicular to the sun's surface.'

So sunspots are indeed places where a very strong magnetic field pokes through the surface of the Sun. This single fact is at the heart of the sunspot phenomenon. Since most people find it difficult to visualize what a magnetic field is and what effects it produces, we need to digress a bit and talk about magnetic fields.

De Magnete

The year 1600 was an eventful one in Europe, starting with Giordano Bruno being burned at the stake for heresy and ending with the formation of the British East India Company, which became a major foundation of the expanding British Empire. The climate in Europe turned cold and damp, leading to an overpopulation of rats and the consequent spread of the plague. And a London physician named

William Gilbert (whose patients included Elizabeth I) published a book, called *De Magnete*, that discussed magnetic bodies and claimed that the Earth itself is a giant magnet. Gilbert died in 1603, apparently from an outbreak of the plague.

Between the Black Sea, the Aegean and the Mediterranean lies Anatolia, or Asia Minor – now largely the modern Republic of Turkey – a crucial cultural crossroad between Asia and Europe going back to Neolithic times. In the Western portion, in Ionia, an ancient Greek tribe named the Magnetes established a large and important commercial and trading city known as Magnesia. It was located a day's travel from Miletus, where the philosopher Thales is credited with discovering that a particular type of stone will attract iron and other stones of the same type, although older reports on this phenomenon from China have been found. In the West, a needle-shaped piece of this stone suspended from

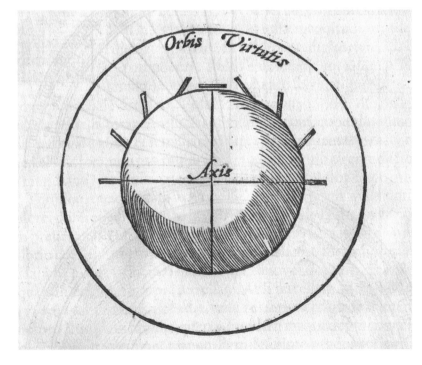

7 Figure from Gilbert's *De Magnete* shows how viewing the Earth as a giant magnet explains the variations in the tilt of a compass needle at different latitudes. Note that the equator is vertical in this drawing and the magnetic poles are at left and right.

a string was used for navigation, as it was found to line up in the north–south direction. From the old English word 'lode', meaning way or course, this path-finding tool came to be called a lodestone.

The strange thing about this lodestone, or magnet, is that it acts at a distance rather than by direct contact. A magnet attracts iron at a distance of at least a few centimetres for weak magnets or much farther for strong ones. The idea developed that the magnet is surrounded by some sort of field, or field of force – which is just another way of saying that it acts at a distance. But the concept of a field has turned out to be extremely fruitful, especially as the properties of the field can be quantified mathematically. For instance, the field surrounding a magnet has a definite shape which can be made visible by sprinkling little bits of iron (iron filings) onto a paper that has a magnet under it, producing the pattern that we've all seen of field lines starting at one end of the magnet – a 'pole' – and arcing around to the other end.

The two poles of a magnet act similarly to the two types of electric charge, opposite poles attracting each other just as opposite charges attract and like poles repelling each other just as like charges repel. So if you bring one magnet near another, its north pole will be attracted to the other one's south pole. As Gilbert demonstrated, the Earth itself acts as a huge magnet with a north and a south pole almost, though not quite, coinciding with the geometric south and north poles. (The magnetic axis of the Earth is tilted by about 10° and slowly moves around. We'll get into that in a later chapter.) The north pole of a magnet is defined as the one that points towards the geometric north pole of the Earth, so the pole that's up north on the Earth is a *south* magnetic pole. The magnet's north pole is attracted to the Earth's south magnetic pole, which is these days located in the north.

Back to Hale's measurement. If the Earth is magnetic, and if there are features on the Sun that have swirls and other structures that make them look as if they are magnetic, then it makes sense to measure to find out if they are indeed magnetic, as Hale did.

What Is a Sunspot?

We are now ready to ask what a sunspot is. We have the beautiful image shown in illustration 2, but what is it that we're seeing?

We can distinguish three main parts to this image. First, there is a dark, roughly circular region known as the *umbra* (Latin for shadow), which is the spot that we can see by eye when we look (carefully and with proper precautions) at the Sun. Surrounding it there is a swirling pattern, reddish-brown in this image, of horizontal thread-like strips known as the *penumbra*. Neither of those words is really appropriate to what we see here, having been developed to explain shadows, but by eye or with a poor telescope, a sunspot does look like the deep shadow and partial shadow that you see in an eclipse. Finally, there is a background region consisting of numerous golden-coloured granules, which are in fact called *granules*. These are the constant background of the un-spotted areas on the Sun representing the normal surface into which the sunspots emerge. They are a type of bubbling motion called convection, typical of a gas or a liquid that is being heated from below. The Sun may look solid, but it is actually a huge ball of dense gas, mainly hydrogen and helium. The best word to describe the Sun is *fluid*, a description that includes liquid, gas and plasma (a liquid or a gas in which the electrons are largely separated from their atoms, allowing the material to become electrically conducting) as possible states.

A digression concerning fluids is in order. In common usage, the word *fluid* is nearly synonymous with liquid. But from the point of view of physics the essential defining property of a fluid is that it will easily deform under applied stress. Water will take on the shape of a container and will flow out of a tipped-over container in response to the pull of gravity, rather than fall in a solid chunk, as an ice cube does.[4] The same sort of deformation will happen, less dramatically, with air when you use a fan, causing the air to flow away from the push of the spinning blades. Gases and plasmas, which deform and flow just as liquids do, are also fluids.

The granules we see on the Sun's surface are flows of fluid, similar to the upwelling and overturning cells that you see in, for instance, a pan of boiling water. The fluid is trying to transfer heat from its bottom surface, where the heat from the flame is being absorbed, to the top surface, where the heat can escape to the air. Conditions in the Sun are similar, where the energy generated by nuclear processes that power the Sun is trying to emerge from deep inside and escape out through the visible surface. Convection – a motion of the fluid – is a highly efficient way of moving heat and it arises frequently in liquids and gases, moving material from hot places to cool places. The Earth's atmosphere, for instance, has large convective circulation cells that bring warm air from the equator up towards the poles, and cooler polar air back down towards the equator. So, like the terrestrial atmosphere, the Sun is not a solid body and it has a variety of motions that a fluid can sustain, but that a solid body cannot.

In the absence of sunspots, the granular convective pattern is seen everywhere on the Sun. The sunspot is an intrusion into this normal surface, and changes it by its presence. These changes are key to explaining what we see in the picture. First, why is the sunspot dark? Careful measurements show that it's because sunspots are cooler than the rest of the solar surface. Cooler is a relative term here: the spots are typically at temperatures of 3,000–4,000 K (kelvins), while the granular surface where sunlight is emitted – the *photosphere* – is at 5,780 K. So the spots are as hot and bright as the small, cool stars known as M-dwarfs, which are inherently (that is, in absolute terms) much fainter than the Sun. The amount of light emitted by a hot body decreases very rapidly as its temperature drops, and M-dwarfs are also smaller than the Sun, so on both

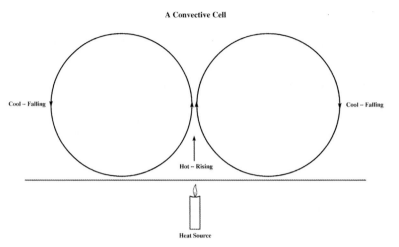

A Convective Cell

Cool – Falling

Cool – Falling

Hot – Rising

Heat Source

8 A heat source below a fluid will heat the fluid locally, causing it to become less dense and therefore to rise, thereby inducing a flow that carries heat upwards. The fluid then recirculates by flowing sideways to get out of the way of the additional rising fluid, then cooling and sinking back down in a heat-transport cycle.

counts they emit far less light. At the typical spot temperatures, the spot brightness is about one-fifth as much as the photosphere's. If the whole Sun were that bright, we would still see it easily, but it would shine a dull red colour and the Earth would be terribly cold. The spot only looks dark by comparison with what surrounds it, because the eye or the camera adjusts to the average brightness level.

But this is still only half an answer: sunspots are dark because they're (relatively) cool, but why are they cool? To answer that, we need to go back to the nature of magnetic fields and to the Danish scientist Hans Christian Ørsted in the year 1820. Ørsted was a follower of the philosopher Immanuel Kant and his ideas about the unity of nature, and he was a friend of the philosopher Friedrich Schelling, who founded the school of Naturphilosophie based on the belief that all of nature should be derivable from a single first principle. Ørsted was interested in electricity and magnetism and was looking for a way to unify them when, during a lecture, he noticed that an electric current in a wire caused a nearby magnetic compass needle to be deflected, with the needle pointing at right angles to the direction along the wire. He decided to pursue this strange effect, and after detailed experimentation he established that the current flowing through the wire produces a magnetic field oriented *around* the wire, in a plane perpendicular to the direction of the current: think of the field as forming a bull's-eye pattern of concentric circles, and the current as an arrow straight into the centre of the bull's-eye. Somehow, the electric current in the wire has an effect on the space surrounding the wire, and in a perpendicular direction to boot. This discovery caused great excitement, combining as it did two separate fields of study, electrical phenomena and magnetic ones. The discovery inspired others, such as André-Marie Ampère and later Michael Faraday, to carry out further experiments that eventually led to the development of a new field of physics, electromagnetism.

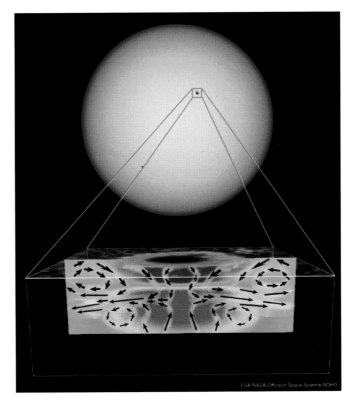

9 The complex interactions between a strong magnetic field bundle that intrudes into the Sun's convective flows and the rising and falling fluid motions near the surface determine the intricate detail of a sunspot. The converging flows just beneath the surface help hold the sunspot together.

The relevance of these discoveries in the laboratory to our discussion is that there is seen to be a symmetry to electric and magnetic behaviours: just as the magnetic needle with its magnetic field is deflected by an electric current, it was also found that if an electric current tries to flow past a magnetic field, it in turn is deflected by the field and forced to circulate around it. Only if the current moves parallel to the magnetic field, instead of trying to cross it, is its direction not deflected. So the charged particles that make up the electric current find it easier to move along the magnetic field than across it, and they end up being constrained to move in that direction by the presence of the field.

This is the situation we have on the Sun in the vicinity of a sunspot. The photosphere with its convective cells is in constant overturning motion, with the hot fluid rising, spreading out like a lava flow and then cooling. The (relatively) cooler fluid then sinks back down at the boundaries between the convective cells, as shown in illustration 8. When the strong magnetic field of the sunspot intrudes into the bubbling, photospheric, conducting plasma, it impedes the convective motions, preventing the hot material from crossing into the bundle of strong magnetic field and instead diverting the flow back out towards the undisturbed field-free surrounding region. This blockage of the free flow of convective material inhibits the ability of the convection to carry heat up to the solar surface, causing less heat to be transported into the magnetized region and making it cooler than the surrounding areas. Hence the spot ends up cooler and looks dark.

Illustration 9 shows an attempt to illustrate part of what is happening below the visible solar surface in the vicinity of a sunspot, although our ability to determine what is going on beneath the surface is still in a crude state. There is a great deal going on in and around the sunspot, involving a large bundle of strong magnetic field – the blue funnel-shaped region just below the surface showing where the sound speed is lower, indicating lower temperature, and the fuzzier and less well-resolved red region further down indicating higher sound speed. This column of (presumably) strong magnetic field protrudes into a bubbling, roiling sea of fluid motions indicated by the directions of the arrows. The appearance of the sunspot, shown in the image at the start of this chapter, is determined by the interaction between the upwelling fluid flow and the intruding magnetic field bundle, and the way in which these processes come to an accommodation.

Measurements of regions such as those shown in illustration 2 show that the bundle of magnetic field starts out being vertical at the centre of the spot in mid-umbra, but the field begins to lose strength and to tilt away from the vertical as we move out from the centre. For large spots, some of the magnetic field at the outer edges of the magnetic field bundle rather suddenly folds down all the way to the surface, forming the iris-like penumbra of the sunspot. The reason for the sharpness of the boundary between umbra and penumbra is still a subject of active discussion, and there is much that we still don't fully understand: is the bundle of magnetic field below the surface a monolithic sheaf, similar to a bouquet of flowers clasped tightly, or do the fields separate into numerous smaller bundles, like the tentacles of a jelly fish – the so-called Medusa or spaghetti model?

Either way, the strength of the magnetic field increases as we move downwards, squeezed into a smaller cross-sectional diameter by the increasing pressure of the Sun as we move farther into the solar interior. The reason the sunspot is cooler is because of this funnel-shaped structure: the heat flowing upwards moves along the bundle of magnetic field and spreads out as it travels upwards because the magnetic structure is increasing in diameter. So the heat is diluted because it is distributed over a larger area and the spot then ends up with less energy density than the surrounding field-free parts of the solar surface.

One small puzzle remains: if the sunspot is a place where there is less power being radiated away than is typical and this power is being diverted away from the bundle of strong magnetic field in the sunspot, there must be some other place where more power than usual is being radiated. In the model shown above, there ought to be a bright ring of extra emission surrounding the sunspot, since it is diverted from the spot area. But no such ring has been observed, despite very careful and sensitive measurements. Why not? The answer seems to be that the ability of those little granules to carry energy is very, very effective, not only in the vertical direction – from inside the Sun up to the surface – but also in the horizontal direction. The excess energy is distributed very efficiently by these bubbling granules as they spread out after having bubbled up, carrying the heat across large areas of the surface around each sunspot. This ends up diluting the excess brightness down to minuscule levels, making the bright ring effect so tiny that it is unobservable.

We end by pointing out something surprising: this chapter began by discussing the visible, naked-eye observations of sunspots. By rights, we ought to have been stuck talking about only what we can see directly, at the surface of the Sun. But somehow, we've also been discussing what is happening beneath the surface, deep down into the Sun. How is this possible? How can we know what is going on down where we can't directly see? That will be the subject of the next chapter.

TWO

Looking Inside the Sun

On 23 August 2011, at 1.51 pm, a magnitude 5.8 earthquake struck the East Coast of the United States, with its epicentre in Virginia. It was not an especially large earthquake – a friend in California wrote to one of us: 'We don't even interrupt a conversation for a 5.8!' But it was unusual for that part of the world and it afforded residents from Georgia to Canada the opportunity to experience the tremors generated by such an event. Within minutes, a brief episode of ground-shaking was felt hundreds or even thousands of miles away. For those paying close attention, there were actually two passing events – first a motion up and down, and then some seconds later another separate event of swaying side to side. In this simple observation lies a wealth of possibilities for understanding the Earth and also the Sun.

Richard Dixon Oldham, a British geologist born in Dublin in 1858, worked for the Geological Survey of India. Although he thought of himself as a geologist, he was best known for his studies of earthquakes in the discipline known as seismology. Working in the Himalayas, a region of massive mountain peaks formed by the crumpling of the Earth's crust from the collision of the Indian tectonic plate with the Eurasian plate, he wrote an important – one might say groundbreaking – report on a large 1897 earthquake in the Assam region in northeast India. In this report he described in careful detail the nature of the buried fault and its relation to the

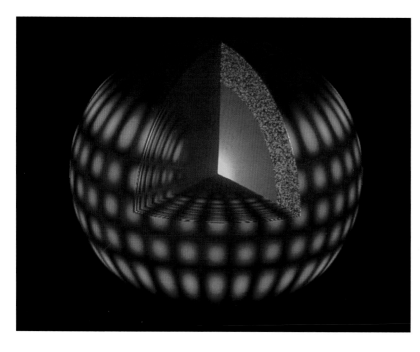

destructive waves that propagated out of the uplift region which had released the enormous amount of energy in the event.

Oldham retired from the service in 1903 and in 1906 published a study of earthquakes in which he detailed his analysis of the waves produced by these events. His analysis revealed two major findings: first, that there are several different types of waves that emanate from an earthquake, and second, that earthquakes detected on the opposite side of the Earth from the source show yet another, delayed, type of wave. From this latter finding he came to the surprising but correct conclusion that the Earth has a core made of some material that transmits waves more slowly than does the overlying and surrounding mantle, and he even made a fairly accurate estimate of the size of this core.

Oldham argued in his 1906 report that a careful study of seismic waves could be used to probe the interior of the Earth:

Just as the spectroscope opened up a new astronomy by enabling the astronomer to determine some of the constituents of which distant stars are composed, so the seismograph, recording the unfelt motion of distant earthquakes, enables us to see into the earth and determine its nature with as great a certainty, up to a certain point, as if we could drive a tunnel through it and take samples of the matter passed through.

There is a great deal packed into this one sentence from his paper, and it will take some discussion to unpack it all. Oldham had already shown in 1900 that the disturbances propagated from an earthquake split into three types of waves: a surface wave, a compressional pressure wave (called a 'P-wave') and a 'distortional' wave (a shear wave, also called an 'S-wave'). The waves that we see on the ocean are surface waves, which travel along the interface between two media having different densities – in this case, water and air. They are the slowest-moving of the three and are also the most destructive. The fastest wave is the compressional wave, caused by pushing on an elastic material, which produces an oscillating compression and rarefaction that moves through the body; this type of wave moves inside the medium, rather than along its outer surface. You may have seen a Slinky doing this if you hold it by one end, let it hang vertically, and move the held end up and down. The third type is the transverse wave, made by shaking the material side to side, that is, by shearing the material. This wave also moves through the medium, progressing via a sideways vibration of the material. One of the notable properties of transverse waves is that they will not propagate through a liquid, because liquids will just flow aside if they are shaken from side to side, so no wave is generated that way.[5]

How did Oldham propose to see inside the Earth? Suppose that we have an array of seismographs spread out over the surface of the Earth – as in fact we do – and that an event occurs somewhere inside

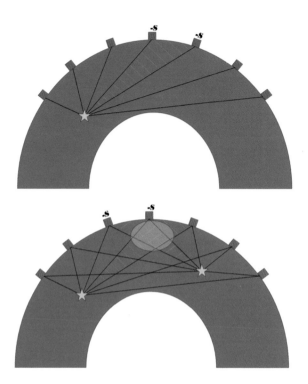

11 An array of seismographs spread across the surface of the Earth can be used to establish where an earthquake occurs and also to work out some properties of the material that the seismic waves pass through on their way to the measuring equipment. In this case, most seismographs measure both the P- and S-waves, but two of the stations (marked '-S') do not record any S-wave, showing that there is buried liquid somewhere in the cross-hatched sector.

12 Adding data from multiple earthquakes located in different parts of the Earth's interior can provide more detailed information about the composition of the interior. Here a second earthquake produces a set of readings at the seismometer network that indicate a different pair of instruments that fails to detect an S-wave. Combining the data from both events helps to localize the underground region that is not allowing shear waves to pass through.

the Earth. The waves propagate in all directions from this event and are detected at the different locations by the different instruments. Because the distances to the seismographs differ, the arrival times will differ – closer ones will tend to see the event sooner – and this connection of time versus distance can be used to triangulate back towards the source of the disturbance and to figure out where the earthquake happened. But more than that can be done. For instance, suppose that all of the stations detect the P-waves, but only some of them detect the S-wave while one sector does not, as shown in illustration 11. Then we know that this sector, indicated by the cross-hatching, contains a large mass of liquid somewhere inside it, because liquids do not allow S-waves to pass through.

If we add a second event from another location to the first (illus. 12), we see that a different sector is picked out as not allowing

S-waves to propagate, that is, as having a liquid composition somewhere in it. Adding the two together, we find a smaller region, consisting of the intersection of the two sectors, which must contain the mass of liquid. There are many earthquakes detected, especially if we use sensitive seismometers that can detect events too small to be felt. These are the ones to which Oldham was referring when he wrote of the 'unfelt motion of distant earthquakes'.

The process we've described is similar to the tomographic imaging methods used in medical CT or MRI scans, where an array of sensors is sequentially moved around an object (your knee, for instance) many times in order to build up a three-dimensional image of the structures inside that object via analysis of all of the individual scans. Each scan yields only a one- or two-dimensional shadow projection of the structure being imaged, but by projecting such scans at numerous angles all the way around the object it is possible via computer analysis to reconstruct the shape of the object being viewed. We've illustrated this method as a way to find liquids, but it can be used for finding many other properties of the structure inside the Earth. Basically any localized feature that alters some wave paths and not others can be imaged in this way. For instance, a large mass of some soft material having a slower sound speed – through which sound moves more slowly – can be found, because the arrival times of waves at some detectors will be later than you would expect them to be based on what the other detectors are showing. You would then conclude that the volume indicated by the oval in illustration 12 has a different composition from its surroundings. Going further, you can determine the speed of sound in that volume and possibly work out what the material is, based on your independent knowledge of sound speeds in different materials. So this sort of analysis allows us to determine what the inside of the Earth is made of, so long as we can get waves to pass through those regions and back out to the surface to be measured and analysed.

Waves on the Sun

Standing at the edge of a swimming pool on a sunny day, preparing
to jump in, you may have noticed a changing network of light and
dark patterns on the bottom of the pool. If you have a propensity
towards looking for explanations, then figuring out where this
pattern comes from is a priority and you may soon realize that the
pattern on the bottom of the pool is due to the ripples of waves
criss-crossing the surface of the water, interacting with and altering
the sunlight entering from above. Essentially, the varying thickness
of the water as the waves form and move across the surface creates
small lenses that focus the light and concentrate it, leading to bright
patches focused onto the bottom of the pool down below. Then,
wondering about the source of these ripples and waves, it is reason-
able to conclude that some sort of disturbance – from strong wind
perhaps or from the people moving around through the water – is
generating them. Reflections from the sides of the pool cause the
waves to bounce back and forth, interacting with each other and
forming the complicated network of ripples visible on the surface.

In addition to the waves seen on the surface of the water, there
are sound waves that propagate *through* the water. If you're a scuba
diver you will have noticed that sound travels through water quite
well. You can also detect this in your bath: if you lean your head
back so your ears are under water (but not your face, since you need
to breathe!) and then scratch the bottom of the tub with your finger-
nail, you will hear the scratching sound very clearly. Sound waves,
which are compressional waves, travel through water as they do
in air; they travel substantially more quickly, in fact, roughly five
times faster.

The sound waves could, in principle, be detected at the surface
of the water when they reach it from underneath and cause a slight
vibration of the surface as they get reflected back down into the
water. (This effect has reportedly been used by spies to listen in

on conversations in a building via detection from a distance of
vibrations of the room's windowpanes.) The relevance of this to
our solar story is that in the early 1960s similar vibrations were
discovered on the surface of the Sun, and they can be used to listen
in on what happens inside the Sun.

Robert B. Leighton was born in Detroit in 1919 and grew up in
downtown Los Angeles after moving there with his mother. Leighton
earned a BS in Electrical Engineering from Caltech in 1941, then
switched to physics and earned an MS in 1944 and a PhD in 1947, both
from Caltech, whose faculty he joined in 1949.[6] All told he spent 58
years at Caltech, carrying out ingenious experiments in solid state
physics, astrophysics, particle physics and radio astronomy, among
other fields. He is probably best known to the public for having
transcribed the Feynman *Lectures on Physics* from recordings of those
lectures, and he also wrote a widely used physics textbook named
Principles of Modern Physics. Among his many areas of interest he
included studies of the Sun, which took up where George Ellery
Hale had left off, revolutionizing the field of solar astrophysics
in the process.

In 1959 Leighton developed the modern method of measuring
magnetic fields on the Sun. Building on the work of Hale and of
his colleague named Horace Babcock, Leighton devised a method
for making very quick measurements over large regions of the
Sun, using a procedure that we now call difference imaging. It is
an extremely useful and sensitive tool for emphasizing places in an
image that differ from the average, by taking pairs of images and
subtracting one from the other. What remains is an image showing
where the difference is between the two, and how large the difference
is. In our digital era this can be done on a computer, but prior to the
final decades of the twentieth century it all had to be done by long,
careful and somewhat tedious photographic methods. Typically,
one image would be turned into a negative via contact photography
(exposing a transparent film image placed directly onto another

13 This image shows a modern Dopplergram (velocity map) from the ground-based Global Oscillation Network Group (GONG). Small patches of the solar surface oscillate up and down with a period of about five minutes, and these motions towards or away from the observer are shown here as, respectively, blue or red patches.

blank transparency); then the second image was placed onto the negative and both were printed through together. If done very carefully, the resulting image is grey where the two original images are the same, and is either bright or dark where a change has occurred in the second image relative to the first, bright for an increase and dark for a decrease.

With this method it became possible to study not just the strong magnetic regions inside the sunspots, but the entire region of activity surrounding spots, with the wavelength shifts from the presence of magnetic fields altering the brightness in the image at the places where the field is located. Beyond that, Leighton realized that the same technique could also be used for measuring motions on the visible solar surface, via the small shifts in wavelength known as Doppler shifts. In this case the wavelength shifts are due to motions either towards or away from the observer, rather than to the presence of a magnetic field, but the difference-image technique is just as applicable to one situation as the other. It is only the way in which we interpret the resulting difference images that differs, in one case showing the strength of the magnetic field and in the other showing the velocity towards or away from the observer.

Using this technique, Leighton and his students Robert Noyes and George Simon discovered that large cells of horizontally moving material cover the solar surface, with material moving upwards in the centres of the cells, horizontally outwards from the centre across the main area of the cell, and downwards at the edges. The surface of the Sun was found to be covered with these large cells and they also found that the magnetic fields measured on the solar surface were formed into a cellular pattern, with the fields tending to be found at the outer edges of the cells. Moreover, they found ubiquitous oscillatory vertical motions (waves) covering the solar surface, in a small-scale pattern and with periods averaging about 300 seconds. The large-scale cells were given the name 'supergranulation', representing as they did a larger version of the

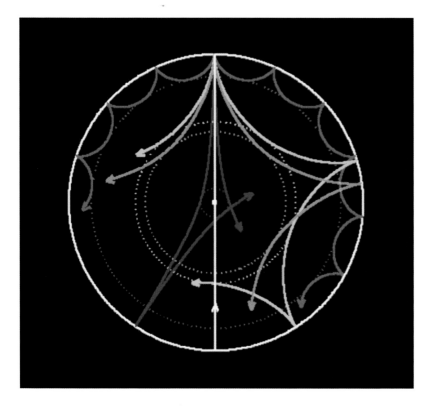

14 P-waves, also called compressional or acoustic waves, travelling inside the Sun may be trapped in a region bounded on top by a large drop in density near the solar surface, and on the bottom by an increase in sound speed that refracts a downwards propagating wave back up towards the surface. The waves return to the surface where they can be detected and analysed to reveal information about the solar interior through which they have passed. As shown in this figure, the long-wavelength waves penetrate more deeply, and these are the long-period, low-frequency waves. If we want to know the conditions deep inside the Sun, we must detect those waves, and this requires very long uninterrupted observations, either from space (HMI) or from a worldwide network of ground-based observing stations, known as the Global Oscillation Network Group (GONG).

well-known small-scale granulation that we saw outside sunspots in images such as illustration 2. The small-scale pattern, called 'five-minute oscillations' because of the 300-second period, was found to be closely related to the granulation.

What nobody knew yet was that this was the start of a new field of study, helioseismology, which would allow us to see inside the Sun. These waves on the Sun can become our array of seismographs, allowing us to measure waves that have reached the surface by travelling through the Sun.

Waves in the Sun

Johannes Kepler (1571–1630) is best known for his three laws
of planetary motion, based on analysis of Tycho Brahe's careful
observations that improved upon the then decades old Copernican
model by demonstrating that the planets orbit the Sun in elliptical
paths, rather than in perfect circles, with the Sun at one focus of
each planet's ellipse. He also sought to ensure that the observations used for his calculations were as accurate as possible and
investigated, among others, the errors in observation produced
by the bending of light from some distant astronomical object as
the light passes through the Earth's atmosphere on its way to the
observer. He concentrated especially on observations of objects
close to the horizon, where the effect is largest, producing detailed
measurements of expected vs observed positions (and he came
within a whisker of deducing the correct sine law for refraction).
What causes refraction is that the speed of propagation of the wave
changes as it moves through the medium. Whether it is sound
waves or light waves propagating, the principle is the same: the
disturbance will bend away from the parts of the medium where
it propagates more quickly, and towards the parts where it moves
more slowly.[7] This bending of the light as it passes through a
dense medium is crucial to the study of the solar interior because
it will allow us to connect waves travelling inside the Sun to waves
observed at the surface.

Ringing Like a Bell

A few years after the five-minute oscillations and the supergranulation
were reported, Roger Ulrich, and independently John Leibacher and
Robert Stein, showed that the surface motions were part of a far more
extensive network of waves, covering the surface and penetrating
deep into the Sun. Of the different types of wave that could exist in

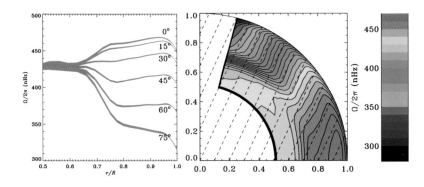

15 A major early triumph of helioseismology was a determination of the fluid motions inside the Sun. The two figures shown here both depict the same result, displayed in two different ways: the equator of the Sun rotates more rapidly than do the poles, and the inside of the Sun rotates rapidly. The graph on the left shows the rotation at selected latitudes, shown from halfway down to the centre of the Sun (0.5 r/R) out to the surface. The vertical scale indicates the rotation rate. The equator, 0°, has the highest rotation rate and latitudes moving up towards the poles (15°, 30°, 45°, etc.) are successively slower. Moving inwards (to the left in this figure), the rotation rates all come together, meeting about 35 per cent in from the surface so that all latitudes are rotating at the same rate. Deep inside the Sun there is a spherical core region that does not rotate differentially at all, but rather as if it is a solid ball. The figure on the right shows this same result displayed in colour contours, with red being the most rapid rotation and blue being the slowest. The interface between the differential rotation and the solid-body rotation is thought to be where the strong magnetic fields are generated. The colour differences disappear deeper than 0.7 r/R, becoming all yellow and gold, indicating that all parts are rotating at about the same speed.

the Sun, the main type of wave that is actually observed is a sound wave, a compressional or pressure wave, designated a 'p-mode'. These waves can move in all directions in the Sun, but waves that reach the top, the photosphere, bounce back down because of the sudden steep drop in density at this boundary. Those that move downwards find themselves moving through a medium with a changing sound speed: the pressure and the density increase towards the inner parts of the Sun, causing an increase in the sound speed. Waves moving inwards bend away from the region of high sound speed, so in this case they are refracted *upwards*, back out of the interior and towards the surface (illus. 14). The downwards- and upwards-moving waves interact with each other, leading to resonances that allow only specific discrete frequencies to survive, as happens on a vibrating string or in a ringing bell. At the surface we can detect the waves that have travelled into the Sun and back out, and compare them to waves that have not made the round trip.

Waves with millions of different frequencies are found, dominated by a spectrum roughly centred on a frequency of five minutes. The beautiful image featured at the start of this chapter depicts one of these wave modes. The five-minute oscillations turn out to be the surface manifestation of the wave modes travelling throughout the Sun. With clever analysis, the swelling oscillations seen at the surface can be made to reveal what the returning waves experienced in their travels through the inside of the Sun, very much in the way Oldham's analysis of seismic waves from earthquakes in India revealed details of the Earth's interior.

One of the major results to come out of this analysis was the depiction of the way that the differential rotation of the Sun – the more rapid rotation of the equator compared to the poles – continues down into the interior of the Sun. It was found that the differential rotation exists only in the outer third of the Sun's volume, the so-called convective zone. At the bottom of the convective zone the Sun was found to have a core, as exists in the Earth: about 30 per cent of the way in towards the centre the differential rotation ends, and beyond that distance the Sun is found to rotate as if it were a solid body, all parts rotating together. This change of rotation pattern is indicated on the plot shown in illustration 15: the rotation curves for different latitudes are widely separated near the surface but come together at an inner radius of about 65 per cent that of the outer radius. Happening as this does at a fairly sharp boundary, there is a large change in rotation rate in a thin layer inside the Sun, as indicated by the closely spaced, nearly horizontal contours at the yellow-gold region in the right part of the figure. This shear layer, as it is called, will play a

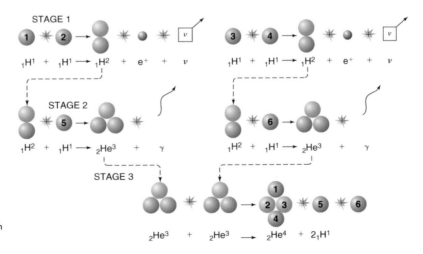

16 Deep inside the Sun hydrogen nuclei (protons, pictured here in red) fuse together under the intense temperature and density at the solar core to form the next heavier element, helium, via a series of steps, several of which involve the decay of a proton into a neutron (shown here in blue) with the release of an electron and a neutrino. ν indicates a neutrino and γ indicates short-wavelength light known as gamma rays.

major role in our next discussion, the generation of strong magnetic fields deep inside the Sun and their eruption to the surface to produce sunspots and the host of related phenomena.

Helioseismology also had a role in helping to resolve the long-standing 'solar neutrino problem'. Deep in the core of the Sun the temperature is so high that hydrogen nuclei – protons – crash into each other at such high speeds that, despite having the same electrical charge and therefore tending to repel each other, they come sufficiently close enough together for them to be able to merge and become a heavier element. Illustration 16 shows a series of such interactions, known as the p-p chain, by which this type of high-speed collision produces a series of heavier nuclei reaching from hydrogen, with one proton, up to helium with four nucleons, two protons and two neutrons. As part of this chain of reactions neutrinos are emitted and make their way out of the Sun, allowing them to be detected from Earth. The problem that only about one-third as many neutrinos were detected as there should have been stubbornly persisted despite years of effort to find an explanation.

Neutrino experiments are notoriously difficult because neutrinos interact so weakly with ordinary matter, so the first place to look for an explanation was the experiment itself. This approach, involving detailed checking of the method and also construction of other experiments using different methods, indicated that the solution lay elsewhere. Another class of experiments involved alteration of the models of the solar interior structure to produce fewer neutrinos. But none of those attempts succeeded, largely because the models, with confirmation about the interior temperature structure of the Sun from helioseismology, could not be altered enough to account for the difference. The indications were that our understanding of neutrino physics had to change, and researchers turned to a 1957 suggestion by Bruno Pontecorvo, who noted that there ought to be three types of neutrinos, and if they are not massless particles like photons but instead have a small mass, then they should spontaneously convert into each other in a type of oscillatory manner. Thus the solar neutrinos, which start out as the type known as electron neutrinos, convert into all three types on their way to the Earth. But the experiments initially in operation were only sensitive to electron neutrinos, so they only detected one-third of the total number! Subsequent experiments were built that allowed this idea to be tested and it turned out to be the correct explanation.

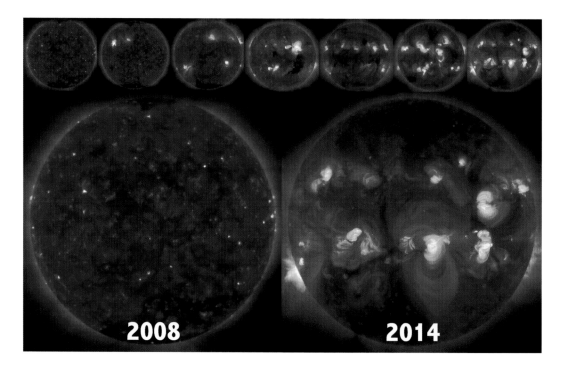

17 The most recent solar cycle, the 24th since sunspot monitoring began in earnest around 1755, peaked in late 2014/early 2015 and then moved into its declining phase. One image per year from the *Hinode* X-ray Telescope (XRT) is displayed above, starting in 2008, just after cycle minimum and marked by the appearance of the first reversed-polarity high-latitude sunspots of the new cycle. Cycle 24 is notable for having the lowest activity levels since the dawn of the space age and the lowest sunspot counts in a century.

THREE

A SOLAR PULSE

William Gilbert's proposed explanation for the Earth's magnetic field in *De Magnete* was that the Earth itself is a giant lodestone, a huge magnet. By way of emphasis, the full title of his famous book included the phrase 'on the Great Magnet the Earth'. It was a plausible idea, based on what was known at the time about magnets, and Gilbert demonstrated how well the idea worked by machining a large lodestone into a spherical shape. This model Earth, a 'terella', reproduced the known behaviour of compass needles as they were moved across the surface, including both the property of pointing towards magnetic north and another more subtle property, the dip of the needle away from the horizontal at different latitudes on the globe.

Alas, Gilbert's nice theory survived only for a single generation. His model predicted that the Earth's magnetic field would be fixed and permanent, unchanging. Not for nothing are they called permanent magnets! It was well known in Gilbert's time that the compass points in a different direction from the actual north. This difference, called the declination, was mapped and plotted so that navigators could make the appropriate corrections, and it could perhaps have been explained in Gilbert's model by saying that the Earth's internal magnetic field is tipped relative to the spin axis of the Earth. But in 1635, a London mathematician named Henry Gellibrand published a work entitled *A Discourse Mathematical of the*

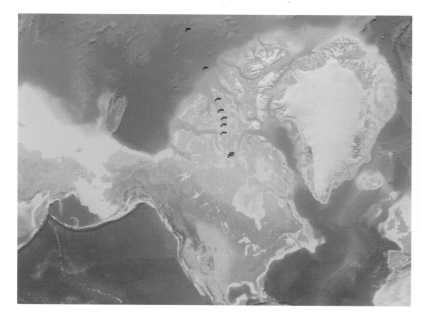

18 The direction of magnetic north measured at any place on the Earth's surface drifts over time. The motion has accelerated recently, with magnetic north moving at a rate of about 15 km/year a century ago to 50 km/year more recently, through the Northwest Passage of Canada. Magnetic north has continued to drift towards the north pole of Earth's spin, though since the date of the northernmost point shown here, the location is no longer detected with on-site expeditions.

Variation of the Magneticall Needle Together with its Admirable Diminution Lately Discovered. This work showed, most surprisingly, that the direction of magnetic north as seen from the vicinity of London was moving, and quite rapidly at that, producing a clearly measureable change in only a few decades. A modern summary of magnetic data is shown in illustration 18, with the somewhat erratic but persistent drift clearly visible. Unless the permanent magnet inside the Earth is somehow moving around, these measurements are very hard to explain.

Just such a model was proposed by the famous mathematician and Astronomer Royal Edmond Halley. He developed a model in which the interior of the Earth contains multiple spherical magnetized shells, moving slightly differently in just the right way to produce the observed drift of the Earth's magnetic field. Halley was so proud of this (incorrect) model that for his official portrait, painted when he turned eighty years old, he chose to hold a drawing of it in his hand. It would be another two centuries before

EDMUNDUS HALLEIUS R.S.S.
Astronomus Regius et Geometriæ Professor Savilianus.

19 Edmond Halley, *c.* 1721, from his *Tabulæ astronomicæ*.

the modern theory explaining how magnetic fields are generated inside the Earth was developed, and then applied to other planets and to the Sun as well.

Cycles on the Sun

The marvellous complexity of sunspots, their effect on their surroundings and on the Earth, and their growth, development and disappearance are only a small part of their story. The

near-vanishing of sunspots for nearly a century, just as they had begun
to be studied, delayed the discovery of a major fact about them.

In 1826 in the German town of Dessau, located about 50 km
north of Leipzig, an amateur astronomer named Heinrich Schwabe
decided to record and study the spots on the Sun. Nobody knows
why he did this, although it is often said that he was searching for a
hypothesized inner planet that was closer to the Sun than Mercury.[8]
The method involves looking for a transit of the planet across the
face of the Sun, a time when the planet lines up between us and
the Sun so that it would look like a small, dark dot sweeping across
the solar disc. In order to detect the planet this way, one would have
to rule out all of the dark dots that are not transiting planets, that
is, the sunspots. Whatever his motivation was, Schwabe made
regular observations of the Sun, day after day, year after year,
recording the numbers and locations of all the visible spots.

Schwabe never did find an inner planet. Indeed, the search for
Vulcan, as it was called, went on for another fifty years and involved
many of the biggest names in astronomy of that era before it was
finally abandoned. But Schwabe did amass a lengthy and detailed
record of the solar spots, and he discovered that they come and
go in a regular cyclical fashion. Awarding the medal of the Royal
Astronomical Society to Schwabe in 1857, the president, M. J.
Johnson, commented:

> It was in 1826 that he entered upon those researches which are
> now to engage our attention . . . [But] it was not until 1843, when
> he had passed through two periods of maximum and minimum,
> that he very modestly remarks that his observations heretofore
> had given indications of periodicity which that year's result
> tended to confirm. Still the subject attracted little attention . . .
> he went on accumulating fresh proofs of his great discovery,
> which, when announced in 1851, by Alexander Von Humboldt in
> the third volume of his immortal *Cosmos*, came upon the world

with all the freshness of novelty, though in reality the secret had been revealed eight years before.

After noting that other terrestrial effects, such as the magnitude of magnetic disturbances, vary in phase with the sunspot number, Johnson concluded with a description of what we would now call the heliosphere, the extent of the Sun's influence throughout the solar system:

> No longer is its scope confined to the disclosure of a physical peculiarity in the constitution of the sun. It promises to be the means of revealing the prevalence of a principle, throughout the solar system, co-extensive with gravitation, and of establishing another link in the chain binding Earth with other worlds.

The point he was making, that the Sun could influence the Earth in some way beyond the known effects of light and gravity, was very controversial at the time. It would be more than a century before the question was finally settled, and the connection between sunspots and terrestrial disturbances was finally understood and accepted.

But first it was necessary to improve our understanding of sunspots. Schwabe had demonstrated that there is a cycle to the comings and goings of sunspots, as shown up to recent times in illustration 20. Suddenly the study of sunspots became one of the hottest areas of research in all of astronomy, and the nineteenth century saw many of the most skilled and innovative observers bring their talents to bear on observing the Sun in new ways and on analysing the wealth of observational data being obtained. Over the next hundred years a wealth of properties characterizing the sunspot cycle were discovered. Among the most prominent are:

1   The rise and fall in the number of sunspots, the time between successive minima being about eleven years, with variations

of two years or so in either direction (illus. 20, bottom part). The amplitudes of the cycles, that is, the numbers of spots appearing at the maximum of each cycle, are highly variable, with occasional periods of very low sunspot numbers occurring.

2   The 'butterfly diagram' for sunspots. Over the course of each cycle, the sunspots at first emerge at high latitudes, far from the solar equator (illus. 20, top part). As the cycle progresses, the emerging sunspots appear closer and closer to the equator, until the new cycle starts with spots once again starting to appear at high latitudes.

3   Magnetic reversal. When magnetic field measurements became available, it was found that sunspots tend to emerge with an east–west orientation, that is, with the magnetic fields of the sunspots oriented mainly in the horizontal direction, nearly parallel to the equator.[9] In a given cycle, most of the spots in

20 Observations of sunspots over many decades show two clear patterns that any theory of their origin must explain. The bottom row shows the area of sunspots on the Sun over the past 150 years; a cyclical behaviour is clearly evident. The top half of the figure shows where on the Sun those spots appeared in latitude, indicating that sunspot emergence begins at high latitudes early in a cycle and then migrates towards the equator as the cycle progresses.

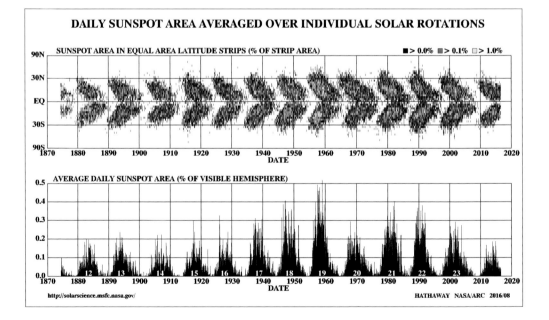

21 A map of the magnetic fields on the solar surface shows that the regions of strong field that produce sunspots tend to be oriented similarly in each hemisphere, with one particular leading polarity favoured; here it is black in the north and white in the south. The polarities are opposite in the two opposite hemispheres, as shown in this magnetic map from around the time of the most recent solar cycle maximum, 20 April 2012. In successive eleven-year sunspot cycles, the orientations are reversed, so that a full magnetic cycle is 22 years.

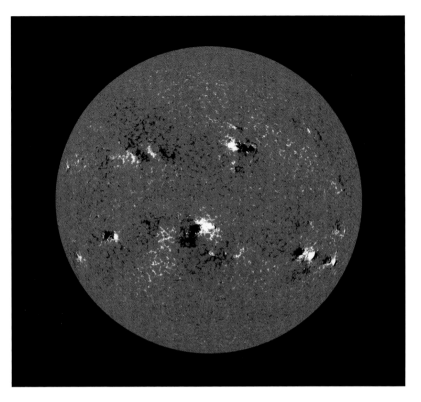

each hemisphere have the same magnetic pattern, with the lead spot (that is, leading in the direction of the Sun's rotation) having a particular magnetic polarity, while spots in the other hemisphere have the opposite magnetic polarity (illus. 21). Discovered by Hale, this rule is known as 'Hale's Law'. In addition, the subsequent sunspot cycle will also have a Hale's Law, but all of the magnetic polarities are reversed. This means that it takes two sunspot cycles before the pattern fully repeats, and the full magnetic cycle is therefore 22 years. The poles of the Sun also switch polarities, with their reversal coming typically around the time of sunspot maximum.

A Heuristic Dynamo Model

When scientists are in the early stages of formulating a theory and have only a partial understanding of the phenomenon under study, it is often useful to propose a preliminary model as a spur to further work. This type of model may well be incomplete, and is certainly provisional, but one hopes that it will at least capture the most essential features, in the form of what might be called an educated guess. Such a model for the solar cycle was put forward in the 1960s, first by Horace W. Babcock in 1961, and then by Robert Leighton in 1969; it is most often referred to as the Babcock–Leighton dynamo model. The key to this heuristic model is that it offers a way to explain the solar cycle using the observed properties of the Sun, especially the differential rotation and the convective patterns of the surface.

As the starting point of the model, we begin with the simplest initial state of the magnetic field, oriented vertically and running from pole to pole, like the field of a bar magnet; this is called a 'poloidal field'. In the model, the field penetrates into the solar interior, which is rotating more rapidly than the surface because of differential rotation. The portions of the field inside the Sun are swept around more rapidly than are the outer portions near the surface, becoming wound up with a horizontal component, as shown in illustration 22. That field, directed around the Sun, is called a 'toroidal field' – one running horizontally, wrapped around the equator like a torus. With this as a starting point the model proposes a series of steps as a way to explain the solar cycle:

1 We start with the simplest magnetic field, a dipole like that of a vertically oriented bar magnet. It will be a poloidal field, with field lines emerging at the top out of one magnetic pole, stretching around the Sun and re-entering near the bottom at the other pole. The field penetrates the Sun and the more rapid

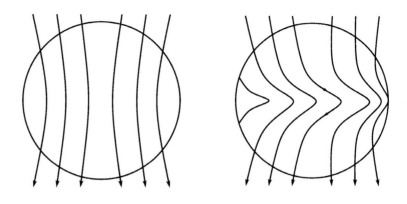

22 The differential rotation of the Sun (left), with the equator rotating more rapidly than the poles (right), stretches out an initial dipole (poloidal) magnetic field so that it develops a toroidal component wrapped around the Sun.

rotation of the solar interior near the equator pulls the field along with it, stretching it out so that it wraps around the Sun (illus. 22).[10] The shear layer at the base of the Sun's convection zone, discussed in the last chapter, is thought to be where most of this stretching out and amplification occurs. This step in the process converts some of the poloidal magnetic field to toroidal field.

2   When the magnetic field inside the Sun becomes strong enough, it develops instabilities that cause the field to erupt upwards, like kinks in a twisted rubber band.

3   The rising loops of magnetic field float up through layers of the Sun which become less dense towards the surface, allowing the field to expand as the local gas pressure of the surrounding solar gas drops. Because of the solar rotation, the rising and expanding field experiences a Coriolis force, similar to the cause of terrestrial cyclones. In the terrestrial case the cyclonic motion develops because air moving into a low-pressure system and moving up from the equator is travelling faster than the air at higher latitudes, while air moving down from high latitudes is travelling more slowly than the air it is moving into. The result is a rotation of the system caused by the inflow of air. For a high-pressure system the air flow is in the opposite direction,

outwards, and the circulation goes in the opposite direction. This is the situation in the Sun, with the rising magnetic field expanding as it moves upwards. The loop of magnetic field slowly rotates as it rises (illus. 23), emerging at the solar surface with a tilt. This tilt has the effect of converting some of the toroidal field back into a poloidal field, but directed oppositely to the poloidal field that originally started the magnetic cycle.

4 After emerging at the solar surface, the magnetic field spreads out across the surface. Because of the tilt with which they emerge, the leading magnetic polarities of each sunspot region are closer to the equator than are the following polarities, so they can interact with and cancel the leading polarities from regions in the other hemisphere across the equator (illus. 24). Babcock had taken the spreading out of the emerged magnetic fields as an observational given, but Leighton contributed a mechanism for it. In the previous chapter we noted that, in addition to the small-scale five-minute vertical oscillations, he had also found a longer-lived pattern of large, cellular horizontal motions and called them supergranulation (illus. 25). Leighton calculated that these supergranules would shuffle the surface fields around

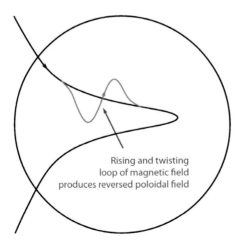

Rising and twisting
loop of magnetic field
produces reversed poloidal field

23 After the differential rotation of the Sun converts a poloidal magnetic field into a toroidal field wrapped around the Sun horizontally, the field becomes unstable and emerges from inside the Sun. As the field rises and expands, the Coriolis force from the solar rotation turns the toroidal field back into a poloidal field, but with the direction reversed from that of the starting field.

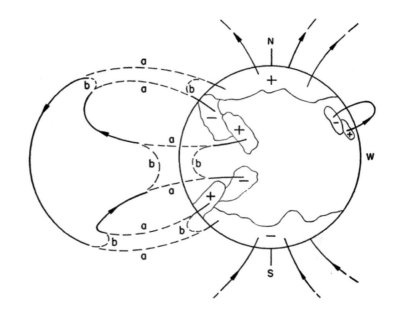

24 The Babcock model of the solar cycle was based on observation of the surface magnetic fields on the Sun and sought to account for the many cycle-related phenomena, such as Hale's polarity law, Joy's law of sunspot tilt and the reversal of magnetic fields in successive cycles. This figure from his 1961 paper shows how the magnetic fields spreading out from sunspot regions cancel across the equator and also cancel the polar fields, thereby reversing the dipole and starting a new cycle.

in a random-walk pattern, sometimes called a 'drunkard's walk' – taking steps in random directions, tending to lead farther and farther away from the starting point. The process, known as turbulent diffusion, has the result that any region of concentrated field, such as that around a sunspot, would spread outwards at a rate given by the size and frequency of the drunkard's steps, and this rate turned out to match what was needed to produce the cancellation of opposite polarity fields.

5 Again because of the tilt, the following or trailing polarities of each sunspot region are farther from the equator and closer to their respective north or south poles and diffuse out towards those poles. Having the opposite polarity of that hemisphere's polar magnetic field, they cancel the field and eventually reverse it. Although this may sound somewhat implausible, it is actually observed. A map showing the evolution of the magnetic field on the Sun over the course of several solar cycles shows these

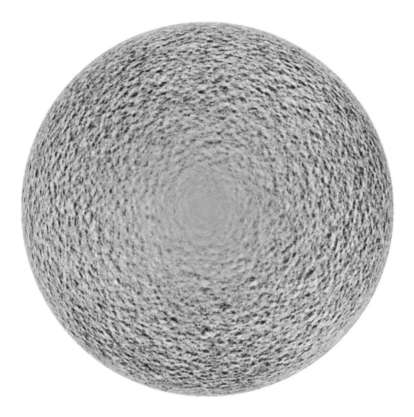

25 This image shows a modern version of Leighton's supergranule image, a map of velocities towards or away from the observer, from the Solar Heliospheric Observatory (SOHO) Michelson Doppler Imager (MDI) instrument, showing supergranules covering the solar surface. The motions of supergranules are parallel to the surface, so there is nearly no component towards us at the centre of the disc. Supergranules move along the Sun's surface, parallel to it; therefore, at the centre of the Sun's disc all their motion is side to side as we see it, with no component of the motion coming towards us. The flow is horizontal, outwards from their centres, with a typical velocity of 400 m per second, and these motions are seen as light and dark crater-like circular features all over the Sun.

opposite-polarity fields travelling towards the poles from the active region latitudes and reversing the magnetic polarities there around the time of sunspot maximum in each cycle (illus. 26).

6   The sunspot cycle then starts over, with all of the magnetic polarities reversed. Illustration 27 shows the rise and fall of solar activity over the course of nearly two sunspot cycles, making up one complete magnetic cycle.

Dynamos Mathematized

The modern explanation for the magnetism of the Sun arose
out of the effort to understand the magnetic field of the Earth,
developed by a shy and complicated scientist who, according to
the solar physicist Eugene Parker (who continued and extended
the work), 'had a deep aversion to asserting himself in the generally
cynical scientific arena'. Walter M. Elsasser was born in 1904 in
Mannheim, Germany, to a prosperous Protestant family that had
converted from Judaism. Elsasser was unaware of his family history
until his teenage years, but it affected his life and career in major
ways as Germany changed in the 1920s and '30s. His first indication
of trouble may have come when his father urged him to apply for
membership in a high-school fraternity in order to help him become
less of a 'nerd'. He was surprised to find himself rejected because
persons of Jewish descent were not permitted into fraternities.

But more serious trouble was years in the future and Elsasser
was able to pursue a wide range of interests, including a strongly
philosophical approach to the field of natural sciences. He came
to conclude that acceptance of scientific ideas depends on whether
they fit into prevailing conceptions, the latter remaining largely
controlled by the unconscious. The family doctor at the time described
him as 'a bundle of nerves', a potential weakness that Elsasser
decided to turn into a strength after reading Marcel Proust's massive

26 The solar surface magnetic field evolution over four sunspot cycles is shown in this map displaying the progression towards the equator of emerging fields throughout each cycle, the drift towards the poles of remnant magnetic field, reversing the polarities of the polar fields, and the overall reversal of north vs south hemispheric magnetic polarities from one cycle to the next.

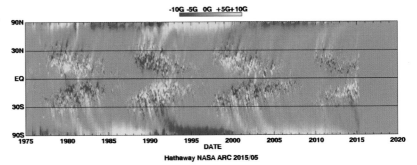

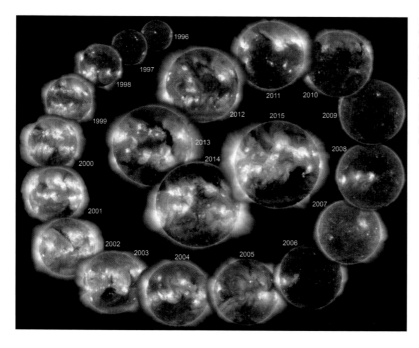

27 This figure shows twenty years of solar activity, nearly two sunspot cycles, as seen by SOHO's Extreme ultraviolet Imaging Telescope (EIT), from solar minimum in 1996, to a maximum in 2001, followed by another minimum in 2009, and then the most recent maximum in 2014/15.

Remembrance of Things Past and thereby beginning to understand how to sublimate neuroses into constructive directions. After graduating from the gymnasium (high school) in 1922 Elsasser went to Heidelberg where, to his dismay, the leading lecturer, a Nobel Prize winner, showed up wearing a large silver swastika. Several people advised Elsasser to leave, and in 1923 he moved to Munich, studying experimental physics with Wilhelm Wien and theory with Arnold Sommerfeld. Although he was happy there, a faculty member eventually pointed out to Elsasser that nearly everyone else on the faculty was a member of the Nazi Party, and advised him to transfer to Göttingen. He did so in 1925, arriving with a letter of introduction to James Franck, who accepted him and within the year encouraged him to publish a short, elegant piece of work explaining some puzzling experimental results on electron scattering from platinum as due to a wave-like behaviour of matter that had been proposed by Einstein and by Louis de Broglie.

Elsasser got his PhD in 1927 and received an unexpected offer from the well-known theorist Paul Ehrenfest to be his assistant in the Netherlands, despite the fact that the offer came with a long letter from Ehrenfest discussing his (Ehrenfest's!) psychological problems. This should have been a warning, because although Elsasser loved the Netherlands, Ehrenfest turned out to treat him at first aggressively and then with outright hostility, for no apparent reason. Eventually Ehrenfest told Elsasser to leave and go back to Berlin, where he had to return to living with his parents. (Ehrenfest committed suicide a few years thereafter.) A university position in Germany was not possible for him in those years and in 1929 he accepted an offer to work at the Kharkov Institute in the Soviet Union. But illness forced him back to Germany, and he ended up in Frankfurt in 1931. In April 1933 the Nazis seized power and Elsasser's psychoanalyst advised him to get to Switzerland before the border closed, which he did after an encounter with some of Hitler's brownshirts, who had occupied the university.

Arriving in Zurich, he was warmly greeted by the eminent theorist Wolfgang Pauli, who knew of a position in Paris, where Frédéric Joliot (son-in-law of Madame Curie) obtained for him a fellowship from the Alliance Israélite Universelle. Elsasser viewed this as an act of charity and was happier the next year when he got a position with the French CNRS (Centre National de la Recherche Scientifique), and he was also then able to help find positions for the numerous other refugee scientists fleeing Germany. His work in those years involved significant contributions to the understanding of the atomic nucleus, work that was completed many years later by J. Hans D. Jensen and Maria Goeppert Mayer, who received a Nobel Prize for it.

In order to stay in France, he would have had to become a citizen and he chose instead to apply for entry into the United States, which was granted in 1935. Although he met his future wife on the boat crossing to the U.S., that was the main positive aspect of the trip as he was unable to find a position anywhere and returned to Paris. A year later he tried again, eventually ending up at Caltech, where the only position available was in geophysics, in the new Meteorology Department, studying the radiative heating and cooling properties of the atmosphere. That position ended abruptly in 1941 when he was unjustly accused of having a 'Washington bureaucrat' (a famous scientist named Carl-Gustav Rossby) exercise undue influence to get a better position for him. It was a false accusation, but rather than defend himself he just packed up and left, ending up at the Blue Hill Observatory in South Boston until the 1941 Pearl Harbor attack led to his being summoned to report for duty by the U.S. Signal Corps. By the end of the war he was working for the Radio Propagation Committee in the Empire State Building, using the weekends to work on his theory of the Earth's magnetism.

He developed the theory in a series of three publications in *Physical Review* during 1946–7, and published a systematic summary of the physics of the Earth's interior in *Reviews of Modern Physics* in 1950. This work was truly pioneering in the sense that nothing like it existed at the time.[11] After first examining the evidence for a liquid iron outer core inside the Earth, along the lines of the studies that Richard Dixon Oldham had pioneered, he also noted that the geological evidence indicated that the direction of the Earth's magnetic field had flipped, with north and south poles changing places, at intervals of roughly several hundred thousand years. Thus not only was a mechanism needed that would generate the Earth's field, but it had to be one that allowed for periodic reversals of the field direction.

Elsasser carried out lengthy calculations showing that a mathematical theory based on turbulent fluid motions of the Earth's liquid iron core (albeit at a speed of only 0.03 cm/year) combined with the rotation of the Earth was the best explanation of the observed properties of the Earth's field. He suggested that the cyclic behaviour of the fields arose out of a feedback between what he termed a poloidal (directed from pole-to-pole inside the Earth) and a toroidal (directed around the Earth, like the equator) magnetic field. As we saw with the Babcock–Leighton model, the same type of theory was later used to explain how the Sun's magnetic field is generated.

Elsasser's theory was mostly ignored, or else strongly opposed by the few who did discuss it, until 1950 when the British mathematician G. K. Batchelor showed that random turbulent motions of a conducting fluid can indeed amplify any slightest stray amount of magnetic field. Thereafter Elsasser's type of dynamo theory was accepted and developed, but by then he had moved on to the work of his final decades, his biological studies centred on a general theory of organisms.

28 Nineteenth-century stained-glass window from the observatory of the pioneer English astronomers William and Margaret Huggins, now on display at Wellesley College, Massachusetts. Included in the stained glass is Fraunhofer's spectrum with prominent absorption lines labelled, three emission lines from a gaseous nebula, a spiral nebula (since discovered to be a spiral galaxy), a comet, the Sun (showing its red prominences and a white corona) and some stars.

FOUR

A Spectrum and What It Tells Us

Among the many miracles that we tend to take for granted is the miracle of vision. Something called light is produced, somehow, and it allows us to see material objects in the world, most of which have some sort of colour.

The word 'light' is used pervasively as a metaphor, in religious thought ('Let there be light'), in describing culture (Dark Ages, Enlightenment), in psychology (a dark mood, a bright person) and as synonymous with truth (to see the light). But behind these symbolic meanings there is a real physical entity called light, and if we try to ask what light is, we begin to find it extremely puzzling. Is it something material? If so, why then is it not possible to hold onto it, to put it in a jar and carry it around? Why don't we feel the impact of light when it hits us? Why doesn't it seem to have any weight if it's material?

What of colour? That appears to be a property of objects, and the light allows us to see the colours. Is the colour there when the light is not shining on it? But if light is made to pass through a piece of coloured glass onto a white surface, then the surface becomes the colour of the glass. Is colour separate from light or is colour in the light? If the latter, why don't we see a coloured beam when the light passes through the air? How can light have colour in it? What would that even mean? And how do we know that light moves from one place to another, instead of, say, being some sort of excitation of what is already there?

Questions about the nature of light have been part of our intellectual heritage for thousands of years, and the story of how we came to understand what light is and how it interacts with matter encompasses nearly the entire history of scientific thought. The process contributed in major ways to teaching us how to differentiate between appearances – how things seem to us after being filtered and processed through our sensory apparatus – and what we might call objective reality, the way things are in the world independent of our perception or even of our existence.[12]

If we are to understand what light is, we need to find a way to distinguish between our perception of light and the characteristics of light that are independent of our perception: does colour per se exist in the world, or is it something produced in our perception by our sensory apparatus? For the scientist, the process began in the seventeenth century with two of the giants of science, Isaac Newton and Christiaan Huygens.[13] The psychological theory, which is strongly rooted in human perception, can be traced to Johann Wolfgang von Goethe (the author of *Faust*) and his 1810 publication *Zur Farbenlehre* (Theory of Colours).

Light and Colour

> Nature and Nature's laws lay hid in night: God said,
> 'Let Newton be!' and all was light.
> – Alexander Pope

The couplet quoted above is, of course, an exaggeration, but gives us some feeling for how Newton's achievements were seen in England after his death. (And eventually, in 1930, Sir John Collings Squire wrote a rejoinder: 'It did not last, the devil shouting "Ho."/ "Let Einstein be," restored the status quo.')

Newton is mainly remembered as a mathematician, the Lucasian Professor of Mathematics at Cambridge, and for his work in

30 Recreation of the discovery by Newton that (left photo) sunlight (white light) entering a prism can be broken down by the prism into a rainbow, and then (right photo) reconstituted into white light by a second, inverted prism. In this recreation at the Huntington Library, San Marino, California, a lever moved the second prism in the bottom photo so that it intercepted the rainbow of colour, refracted it back to the right and recombined the colours into white light.

Fig. 18.

29 Newton's *experimentum crucis*, drawn much later for reproduction in his book *Optics* (1704). Sunlight enters from the right.

developing the calculus, the laws of motion in physics and the theory of universal gravitation that helped explain the motions of the planets. But he was also an outstanding experimenter, constructing the first known reflecting telescope (using mirrors instead of lenses to focus the incoming light) and carrying out a long series of careful experiments analysing the nature of light and of colour. He spent his later years as Master of the Royal Mint, helping to convict and hang dozens of counterfeiters.

Newton and Colour

Newton was born on Christmas Day of 1642 on the Julian calendar then in use (though when transformed to our current Gregorian

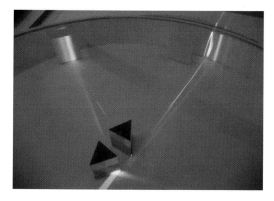
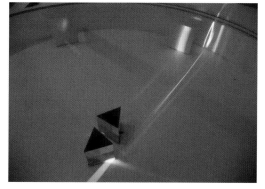

calendar, he was born on 4 January 1643). Young Isaac Newton, soon after taking his bachelor's degree from Trinity College, Cambridge, was among those sent home from university because of the plague in 1665–7. While there, in Woolsthorpe, he carried out a series of experiments on the nature of light.

In Newton's *experimentum crucis* ('crucial experiment', as it is still known), Newton tested one of the prevailing hypotheses of the time: that in the process of refracting (bending) the light, a prism produces the colours that are observed in the light beam exiting the prism. Newton sent a shaft of sunlight from a hole in his window shutter to hit a prism. That action broke up the sunlight into a rainbow of colours. He then isolated one of the colours using a small hole in a second screen and sent it through another prism, finding that it was not split into further colours and that it refracted (that is, bent) the same amount the second time as it had the first. Newton concluded that the refraction does not cause the colours but rather that light consists of 'difform rays' differing in their 'degree of refrangibility', the refraction making them visible but not causing them.

In another experiment, Newton used a lens to image the rainbow of colours from the first prism onto another prism, held upside-down. That second prism combined the colours into white light again (illus. 30). And that new shaft of white light was sent through a third prism, right-side-up, which made a rainbow again, this time projected onto a screen.

To the five colours that Newton originally considered, he added indigo and orange, to make seven colours matching the seven tones on a musical scale. We still often recall the ordering of the colours with the mnemonic matching an apparent name: ROY G BIV, for red, orange, yellow, green, blue, indigo and violet. Newton may have been one of the rare people who distinguish indigo as a separate colour, while most of us see it as a shade of blue.

Newton became aware of the 1678 theory of Christiaan Huygens (1629–1695) that light was made of waves, not particles. After all, two light beams could go through each other without any obvious interaction between the beams. But Newton pointed out problems with Huygens's wave theory: 'If light consists of undulations in an elastic medium it should diverge in every direction from each new centre of disturbance.' Newton and Huygens met in 1689, when Huygens came to London along with the Dutch king ascending the British throne. They subsequently carried on a correspondence about questions of optics and other matters, but the question of waves vs particles was not settled. In Newton's *Optics* of 1704, Newton presented a corpuscular theory of light – that is, a theory that light is made of tiny particles.

The battle between the wave theory and the corpuscular theory went on for centuries. Each type of theory – wave or particle – could explain some, but not all, of the properties of light. The question was eventually resolved only in the quantum-mechanical explanation that light sometimes acts as a wave and sometimes as a set of particles (which we call photons), depending on how it is being studied. We now speak of the 'wave-particle duality'.

Theories of Vision

We know that Plato and Aristotle differed about 2,400 years ago over the question of whether the eye sends out rays that somehow sense or feel the objects that we are looking at or whether it receives rays coming from those objects. Hero of Alexandria in about 100 CE had formulated the 'shortest path' principle, according to which light travelling between two places follows the shortest path; he later added that this is equivalent to following the path requiring the least time. But this principle works no matter which direction – *from* the eye or *to* the eye – the light follows. In many situations light can be thought of as following a ray-like path and we cannot tell which way it is actually travelling, since it moves too quickly for us to see any motion along the path, and both directions are geometrically equivalent. In the late fifteenth century, Leonardo da Vinci first agreed with the Platonic point of view, and then changed his mind to conclude that rays came only into the eye. But if light enters the eye it is not at all obvious how an image of an object is formed, a problem that was discussed with great acuity by the eleventh-century mathematician and scientist Ibn al-Haytham (Alhazen). His work became known to the West via the writings of John Peckham in the thirteenth century, and it was not greatly improved upon until the work of Johannes Kepler six hundred years later.

We respect Kepler for many things, chiefly his discovery that the orbits of the planets (and, by extension, other celestial objects) are elliptical, not simply circular. His first two laws of orbits were published in his *Astronomia nova* (The New Astronomy) in 1609; his third law, governing the speeds of the planets in their orbits as a function of their distance from the Sun, was published in *Harmonices mundi* (Harmony of the World) in 1618.[14]

For our purposes in this book, we discuss other successes of Kepler's fertile mind. He may even have been the first to mention the solar corona, in his 1606 book about the supernova that is still named after him. More generally known is his work on vision. In 1604, Kepler figured out that the eye is an optical device, with an actual image projected upon the back of the eye by a lens located near the pupil. But given the way lenses image, that projected image would be upside-down. Nowadays, we don't have trouble thinking that the human brain automatically inverts the image, allowing us to see things right-side-up, corresponding to their actual orientation.[15] But the idea was controversial in Kepler's time and he devoted a lengthy discussion to presenting a (mostly theological) argument explaining why this inversion makes sense. By 1619 in his *Oculus* and a decade later in his sunspot book, *Rosa ursina*, the Jesuit astronomer Christopher Scheiner agreed with Kepler's imaging idea, but not everybody at the time did. Pierre Gassendi (1592–1655), the first to have seen a transit of Mercury across the face of the sun, was one of those opposed. He and some others thought there had to be a mirror somewhere in the eye to re-invert the image so that it looked right-side-up.

Light as Waves

Huygens's 1678 wave theory of light, involving repeated formation of spreading spherical (or circular in 2D) wavefronts of light as it propagates through a medium, showed light spreading out after it has passed through a narrow slit. A famous experiment showing that light acts as waves is credited to the English polymath Thomas Young in 1803. He shone light through a pair of slits close to each other, so that the light spread out from each slit, with the result that an interference pattern appeared on a screen beyond the slits. This implied that a single wavefront was impinging on the two slits, producing two circularly propagating sources that interact with each other coming out of the slits. Newton's corpuscular theory of light could not explain such interference. In 1815, Augustin Fresnel – now most famous for his design for saving weight and thickness of lenses used in lighthouses – provided mathematical backing for Young's double-slit experiment.

Light as Particles

Nineteenth-century studies of light led to basic laws that astronomers still use. Wien's displacement law shows that the peak of strength of radiation – the wavelength at which the emission of light from a hot object reaches its maximum strength – depends on the temperature of the emitting body, with the wavelength of the peak being inversely proportional to the temperature. The Stefan–Boltzmann law showed that the total energy emitted by a hot body depends on the fourth power of its absolute temperature; thus a doubling of the temperature increases the energy emitted by $2 \times 2 \times 2 \times 2 = 16$ times.

At the turn of the twentieth century, Max Planck, in Germany, provided a formula that explained Wien's displacement law and the Stefan–Boltzmann law. As part of inventing his formula, Planck found that he had to use the mathematical construct of packets of energy, quanta, though he didn't think these bundles were real. It took Albert Einstein, in his miracle year of 1905 when he also presented his special theory of relativity and his explanation of the motions of tiny particles in liquids with their so-called Brownian motion, to give reality to the concept of bundles of energy. He advanced the idea that these quanta of energy make particles of light called photons, and their energy E is inversely linked to the apparent wavelength λ ($E = hc/\lambda$). The equation uses a constant multiplier h known as Planck's constant and c is the speed of light. It was for this work, used to explain the photoelectric effect, that Einstein later received the Nobel Prize.

Following some preliminary work in 1913 by Niels Bohr and then the actual formulation of quantum mechanics by Erwin Schrödinger and Werner Heisenberg in the 1920s, quantum mechanics has become a major governing theory of physics. But Einstein's general theory of relativity of 1915 doesn't include quantum ideas, so we are unable to combine quantum mechanics and general relativity, meaning that these two theories are incomplete. The quest to connect or combine them continues to this day.

Details of the mechanism of how we perceive objects, the interaction of light with matter, are best explained with the later theory of quantum electrodynamics, QED. The subject is described in Richard Feynman's lecture series on the subject, QED: The Strange Theory of Light and Matter, using his diagrammatic way of keeping track of interactions, so-called Feynman diagrams (illus. 31).

31 Feynman diagram showing particles (in this case muon neutrinos), drawn on request by Professor Feynman for one of us (JMP) merely to illustrate what one looks like. We see the idea that two particles (arrows at left and at right) are interacting via a mediating exchanged particle (wavy line in the middle).

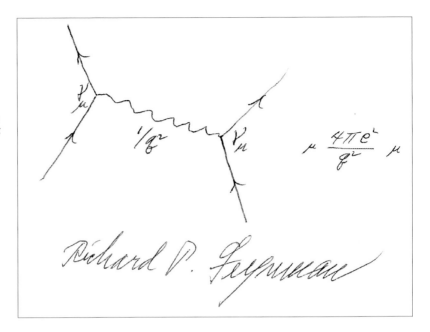

The Spectrum: Light and Beyond

People on Earth have long known of the spectrum, which is displayed to us naturally in the form of the rainbow. Rainbows are located in the sky opposite to the Sun, and form when sunlight is reflected and refracted within raindrops. The philosopher and mathematician René Descartes (1596–1650) was the first to explain the rainbow as due to the internal reflection and refraction of light by raindrops, with the light emerging at a particular angle with respect to the source of the light. The refracted light then emerges from the cloud of droplets in a narrow cone at that angle and forms a large halo around the direction to the source. The different colours are refracted at slightly different angles, as in Newton's prisms, so that they are slightly separated in the halo and form adjacent concentric arcs of the different rainbow colours.

Note that in the main rainbow, the colours are ROY G BIV from the outside in. Sometimes, as in the illustration, there is a secondary

rainbow, which involves an additional reflection inside raindrops. In the secondary bow, the extra bounce causes the colours to emerge in reversed order.

For making a spectrum, reflection and refraction are very different. All wavelengths of light reflect from a surface at the same angle at which they hit the surface: angle of incidence equals angle of reflection. So no colour effects are introduced in a reflection. This is why reflecting mirrors are superior to lenses in telescopes if you are looking at objects that emit a wide range of colours, and you want all of the colours to focus the same way.

But refraction is very different, and is typified by what goes on when white light hits a prism, as in illustration 33. The speed of light in glass or plastic is slower than the speed of light in air (which is itself slightly slower than the speed of light in a vacuum – so-called empty space – which is Einstein's speed c). And the speed of light in a substance depends on the wavelength of the light.

So by the time the light beam goes across the prism, it is bent by an amount that depends on the colour, and it leaves the prism in a rainbow beam.[16]

Our Sun's radiation is strongest in the yellow-green part of the spectrum, and our eyes have no doubt evolved to be most sensitive there. In what we call the 'visual' or 'optical' part of the spectrum, we see the colours from red to violet. Our eyes are insensitive to longer or to shorter wavelengths. Indeed, most of the shorter wavelengths do not even come through the Earth's atmosphere. In Chapter Seven we will see how far into our atmosphere the different wavelengths of radiation from stars and other celestial bodies penetrate before being absorbed. Only in what we call windows of transparency does celestial radiation reach the Earth's surface. There is a window in the visible, a window in the radio part of the spectrum and some narrow windows in the part of the infrared nearest to the visible limit. The human eye contains two kinds of sensing organs. The more sensitive ones are called rods because of their shape. But they perceive only in what we interpret

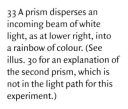

33 A prism disperses an incoming beam of white light, as at lower right, into a rainbow of colour. (See illus. 30 for an explanation of the second prism, which is not in the light path for this experiment.)

as black-and-white. Less sensitive but differentiated across the spectrum are the colour cones.

In nature there is light of different wavelengths, but there is nothing corresponding directly to 'colour'. For us to perceive colour there must be an interaction in the brain among the signals received from the rods and cones in the retina, with colour-differentiated signals being produced by the cones. Humans have three types of cones, roughly sensitive to red, green and blue, respectively. When struck by light, the cones give off a protein called opsin and a molecule called a chromophore. Rods are the black-and-white sensors, producing a higher-resolution but monochromatic image by giving off the pigment rhodopsin. The result of these chemical reactions is that corresponding electrical signals are produced. The retina contains complex connections and circuitry that analyses the signals, taking image contrast and edge detection into account. The processed retinal signals are sent to the visual cortex, where the equally complex task of interpreting the image signal is carried out.

The Solar Spectrum

From the seventeenth-century discovery of the spectrum of sunlight by Newton to the early nineteenth century, there were many advances in optics. But photography was yet to be invented, so the observations were all visual and had to be drawn or described, and much of the ensuing discussion revolved around the difference between the physical properties of light and our perception of light. The two are related, but the struggle to separate objective properties from those that depended on the human perceptual apparatus was long and difficult.

Newton had not seen any differentiations across the spectrum other than a continuous band of colour. But in 1802, William Hyde Wollaston in England reported a division of the visual spectrum of the Sun into four, rather than seven, parts (illus. 34). His main

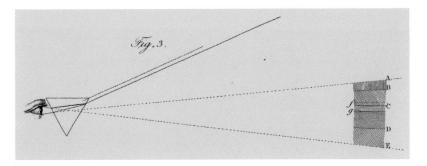

34 In 1802 Wollaston wrote: 'The line A that bounds the red side of the spectrum is somewhat confused, which seems in part owing to want of power in the eye to converge red light. The line B, between red and green, in a certain position of the prism, is perfectly distinct; so also are D and E, the two limits of violet. But C, the limit of green and blue, is not so clearly marked as the rest; and there are also, on each side of this limit, other distinct dark lines, f and g, either of which, in an imperfect experiment, might be mistaken for the boundary of these colours.' (*Philosophical Transactions of the Royal Society of London,* LXXXXII (1802), pp. 365–80.)

advance was to obtain greatly improved observations of the solar spectrum and his discovery of strong, dark absorption lines. The improvement seems to have been due to his using a narrow slit to pass sunlight into the prism, rather than using the whole solar image, which winds up producing such overlapping spectra that the details are smeared out. Wollaston, a doctor who had turned to experimental chemistry and physics, made a fortune by inventing a process – which he kept secret most of his life – for producing malleable platinum from its ore, discovering the element palladium as well. His optical studies were a small part of the work he did, although they formed a large fraction of his published work. Concerning his examination of the solar spectrum, Wollaston wrote, in the 1802 issue of the *Philosophical Transactions of the Royal Society*:

> I cannot conclude these observations on dispersion, without remarking that the colours into which a beam of white light is separable by refraction, appear to me to be neither 7, as they are usually seen in the rainbow, nor reducible by any means (that I can find) to 3, as some persons have conceived; but that, by employing a very narrow pencil of light, 4 primary divisions of the prismatic spectrum may be seen, with a degree of distinctness that, I believe, has not been described nor observed before.

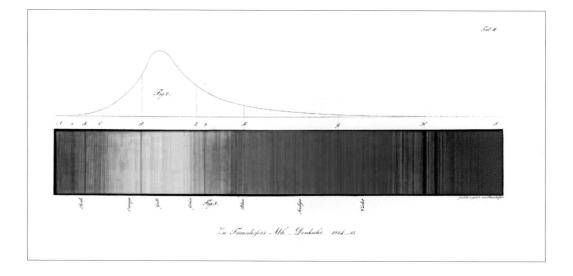

The story continues with Joseph Fraunhofer, who had been orphaned at the age of eleven in 1798. As an apprentice to a glass-maker, he was buried in rubble when the workplace collapsed. In an unbelievable but seemingly true story, he was rescued by a team led by the Prince Elector of Bavaria, who took him under his wing. Eventually, Fraunhofer became employed at a glass works in a former Benedictine monastery, a story well told in episode five of Neil deGrasse Tyson's *Cosmos: A Spacetime Odyssey* (2014), a remake of Carl Sagan's *Cosmos: A Personal Voyage* (1980); this is the same monastery in which the irreverent thirteenth-century manuscript known as *Carmina Burana* was found and used by composer Carl Orff in the twentieth century.

To study the properties of the high-quality glass they were making, Fraunhofer invented measuring instruments, including the spectroscope. From finding a bright yellow-orange line across the spectrum of a flame (now known to be from sodium), he turned the spectroscope on the Sun. To his amazement – and to the benefit of posterity – he found many dark lines across the rainbow of the solar spectrum. These lines are dark – that is, absorbed below the

35 Fraunhofer's original diagram from 1814 (published as a black-and-white engraving in 1817 and colourized here). Above the spectrum he drew a curve of the overall brightness of the solar radiation, showing that it peaked in the yellow and fell off towards both shorter and longer wavelengths.

brightness of the continuum of colour – so they are called absorption lines. We now call them, indeed, Fraunhofer lines (illus. 35).

On Fraunhofer's original diagram, he drew 574 absorption lines. He labelled the strongest with capital letters A through H, and then interspersed lower-case and subscripted letters for the weaker lines. He used capital I to mark the end of the spectrum. We still use his letters for these Fraunhofer lines. The A and B lines we now know are from absorption in the Earth's atmosphere, not from the Sun. The C line is caused by hydrogen in the Sun's photosphere, the visible layer of its atmosphere (that is, the spherical surface from which *photos-*, Greek for light, comes). The D line, actually a closely spaced pair of lines, is from sodium, and can be seen with the eye as bright yellow or orange if you throw salt (NaCl) into a flame. The H line is clearly one of a pair; the French scientist Éleuthère Manscart in 1863 labelled the other one K, though most scientists still mistakenly think it was labelled by Fraunhofer. H and K are the strongest Fraunhofer lines; they are caused by absorption in the solar atmosphere by ionized calcium. Though there is much more hydrogen than calcium in the solar atmosphere, the strongest hydrogen lines are far in the ultraviolet, invisible from Earth's surface because they do not penetrate our atmosphere. A visible C line, known as hydrogen-alpha (Hα), is from a secondary set of hydrogen spectral lines. In recognition of his discoveries, Fraunhofer was ennobled in 1824, which added the 'von' to his name.

We now know, starting even with Fraunhofer's looking at the spectrum of a star, that all stars have Fraunhofer lines. Astronomers use them to interpret the temperatures and pressures of the stars' surfaces. Spectroscopy, starting with Wollaston's and Fraunhofer's work, has transformed the astronomy of the previous millennia into the astrophysics of the twentieth and twenty-first centuries, largely due to the work of the two German scientists Kirchhoff and Bunsen.

On to Astrophysics

Merely seeing dark lines in the spectra of the sun and of other stars doesn't tell you what is causing the lines. In a spectacularly wrong prediction, the philosopher Auguste Comte wrote, in 1842,

> Of all objects, the planets are those which appear to us under the least varied aspect. We see how we may determine their forms, their distances, their bulk, and their motions, but we can never know anything of their chemical or mineralogical structure; and, much less, that of organized beings living on their surface.

He held out even less hope for understanding the stars than for the planets. Soon after making this prediction, it was shown to be dramatically wrong by Bunsen and Kirchhoff.

Gustav Kirchhoff (1824–1887) took the idea of Fraunhofer lines to the next level. At Heidelberg, in Germany, he collaborated with Robert Bunsen (1811–1899) in spectroscopic experiments involving light emitted by hot gases. Kirchhoff and Bunsen identified many spectral lines in the light of the glowing gas with the elements that produce the light, finding that each element has its own unique set of spectral lines, a sort of chemical fingerprint. Almost immediately, using spectroscopy they discovered two new chemical elements, caesium and rubidium, in 1861. For their work, Bunsen used his recently developed eponymous burner, though that was among the least of his accomplishments: he also developed an antidote for arsenic poisoning which saved his life when an arsenic compound exploded in his laboratory, blinding him in one eye and nearly poisoning him. Bunsen also invented the zinc-carbon battery and analysed exhaust gas from industrial furnaces in Germany and the UK to show that they were wasting enormous amounts of energy, leading to revisions that greatly improved the equipment. His famous burner was used to provide a nearly colourless flame for heating the various samples of materials whose spectra he and Kirchhoff wanted to study without interference from any spectral lines of the flame itself.

Kirchhoff figured out how to determine whether spectral lines would appear as absorption (as do the dark Fraunhofer lines in the Sun and stars) or in emission, that is, brighter than their surrounding colours. The appearance turned out to be related to the relative temperature of a background source of radiation and a foreground source, so that what we see is affected by the nature of the intervening material between us and the source of the light. He formulated three laws for spectroscopy to help interpret the observations: 1. a solid (or very dense) hot object produces light with a continuous spectrum; Kirchhoff coined the term 'black-body radiation' to describe the basic, pure shape of the graph of such a thermal source's radiation – the intensity of the light at different wavelengths – for sources at different temperatures; 2. a hot, tenuous gas produces light with spectral lines at discrete wavelengths, with the wavelengths of the lines depending on the chemical composition of the gas; 3. a hot solid object surrounded by a cool tenuous gas produces light with a continuous spectrum that has gaps at discrete wavelengths, those absorption lines corresponding to the emission lines produced by that gas when it is heated. These laws showed how to understand the nature of the sources of the spectra observed in light from distant stars. Kirchhoff also understood, by thinking about the way that hot objects emit and absorb radiation, that a cavity surrounded by hot material has to be filled with radiation corresponding to the temperature of the surrounding material. This discovery put to an end the idea that the inside of the Sun might be hollow and cool enough to support life, given that it is surrounded by such a hot surface.

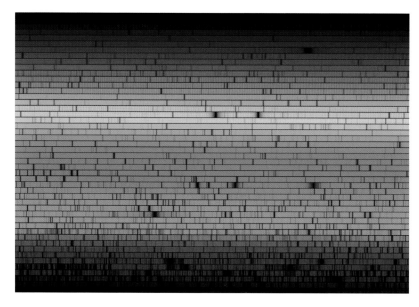

36 Display of a digital recreation of the solar spectrum taken at the National Solar Observatory at Kitt Peak, Arizona, spread out so much that many Fraunhofer lines show clearly.

Interpreting the Solar Spectrum

Calling something a 'spectroscope' implies that you are looking through the instrument with your eye; common use is 'spectrograph' for a photographic recording and 'spectrometer' for an electronic scan. Today's spectrographs are much more capable than Fraunhofer's or Kirchhoff's, of course. Using modern instruments, we now have catalogues of about a million spectral lines in the Fraunhofer spectrum of the Sun. Illustration 36 displays the visible light solar spectrum, recreated from a digital process that provided very high spectral resolution on the Sun.

Each of the spectral lines comes from a transition in energy level of an electron in an atom in the solar atmosphere. A prominent, strong line named Hα, for example, comes from a transition from level 2 to level 3 of the hydrogen atom. The British-American astrophysicist Cecilia Payne (later Payne-Gaposchkin) in 1925 worked out from spectroscopy that the Sun is 90 per cent or so hydrogen. Hydrogen has only one electron producing emission lines via transitions between two energy levels, so its spectrum is very simple, with fairly few spectral lines. Iron, one of the many other elements found in small quantities in the solar atmosphere, has 26 electrons in its neutral state, yielding many more possible pairs of energy levels, so it shows many hundreds of spectral lines in the solar spectrum. Catalogues exist, by Charlotte Moore and others since, of thousands of energy-level transitions from dozens of elements to explain the solar Fraunhofer lines.

In the late nineteenth and early twentieth centuries, a team of computers (that is, people who computed things) at the Harvard College Observatory examined and classified the spectra of thousands of stars. Annie Jump Cannon singlehandedly classified over a hundred thousand stars by the strength of the hydrogen lines in the spectra, with spectral type A being the strongest. She also determined, in the early twentieth century, that hydrogen lines could be weaker either because the star was hotter than the stars of spectral type A or because the star was cooler. So Cannon's alphabetical sequence was reordered by temperature, yielding our now-familiar list of spectral types: O B A F G K M.

37 Display of the spectra of a large range of stars having different surface temperatures, from the hottest at the top to the coolest at the bottom.

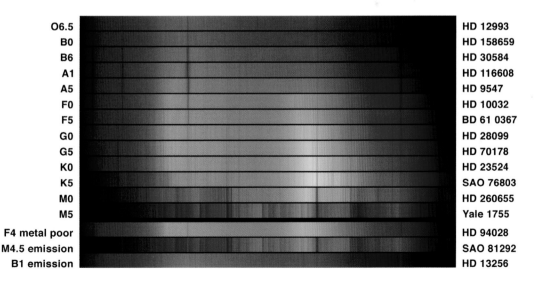

O6.5	HD 12993
B0	HD 158659
B6	HD 30584
A1	HD 116608
A5	HD 9547
F0	HD 10032
F5	BD 61 0367
G0	HD 28099
G5	HD 70178
K0	HD 23524
K5	SAO 76803
M0	HD 260655
M5	Yale 1755
F4 metal poor	HD 94028
M4.5 emission	SAO 81292
B1 emission	HD 13256

With advances into infrared technology, we can now add spectral types for even cooler stars (using previously unused letters): L T Y.

A set of spectra comparing stars of different surface temperatures appears in illustration 37. When elements in these stellar atmospheres become hotter, some of the more weakly held outer electrons in each atom get stripped away, effectively reducing the number of electrons available to form spectral lines. The atoms and molecules of the coolest spectral type stars retain more of their electrons and therefore show more spectral lines. Our Sun is a G2 star, that is, 2/10 of the way between types G and K.

The spectral types are shown at the left in this figure, and the stars' catalogue names are shown at right. Note that the coolest stars shown, of type M, show more spectral lines (many from molecules, which don't survive higher temperatures) than those of hotter surfaces. Looking at spectral types between B0 and G0, the hydrogen series showing a red line, a green line and a blue line is especially easy to see, appearing strongest in spectral type A.

A hundred years ago, the Princeton University astronomer Henry Norris Russell looked at some clusters of stars in the sky and noticed a correlation between the temperatures of the stars and their intrinsic brightnesses. By plotting the location on a graph of each star's temperature and brightness, he found that most of the stars followed a slanted line, which he called the main sequence, across the graph. He also found some stars that were especially bright compared with main-sequence stars of the same colour. Since stars of the same colour have the same temperature and therefore the same brightness per unit of surface area, he concluded that they must be larger. These stars at the upper right of what was long called the Russell diagram must be very large, so he called them giants. In comparison, the normal, main-sequence stars are known as dwarfs. The Sun is such a dwarf star, of spectral type G2.

The Danish astronomer Ejnar Hertzsprung had earlier plotted some similar graphs but he did it for the Pleiades, a young cluster

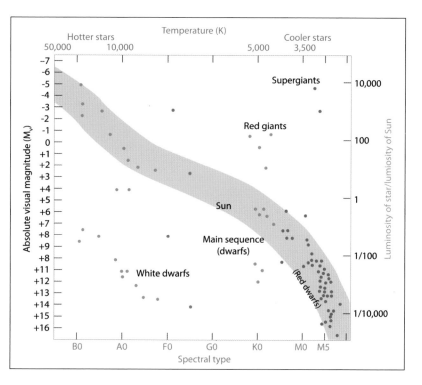

38 The so-called Hertzsprung–Russell, or H–R, diagram plots the absolute luminosity of stars against their temperature. Most stars fall along a diagonal line called the main sequence, sloping down towards fainter and cooler ones. Brighter, hotter stars use up their fuel very quickly and evolve off of the main sequence before the smaller, longer-lived cool stars.

that has only main-sequence stars, which have not had time to evolve into giants. So he missed the key points of Russell's diagram, as well as having published in an obscure German photography journal. But in the 1940s, some other non-American astronomers convinced Subrahmanyan Chandrasekhar, then editor of the *Astrophysical Journal*, to add Hertzsprung's name to the Russell diagram, and Chandra eventually gave in to their badgering. Now it is known as the Hertzsprung–Russell diagram (illus. 38), and is an important tool that astronomers use to plot and study the evolution of stars and star clusters over billions of years.

39 The beginning of totality at the 2013 solar eclipse, viewed from Gabon, showed a reddish rim on the Sun – the solar chromosphere. Also visible is a reddish erupting prominence, material at a temperature similar to that of the chromosphere, as well as two white coronal mass ejections (at middle-right and at far middle-left).

FIVE

THE SOLAR CHROMOSPHERE AND PROMINENCES

The everyday Sun shines brightly on us (when there are no clouds, of course), but what seems to be the sharp edge of the bright circular solar disc isn't really the edge of the Sun. As we have seen, the solar atmosphere continues above the 'everyday Sun', our colloquial name for the solar photosphere. Just above the visible surface there is a region where the temperature starts to go up, instead of going down as it might be expected to. Once treated by theoreticians as a smooth layer surrounding the photosphere, it is actually a complicated region made of tiny spikes of material – if features that are thousands of kilometres long and thus larger than the United Kingdom could be called 'tiny'. They are only tiny compared to the size of the Sun itself. This spiky part of the solar atmosphere is called the chromosphere. In it, the temperature rises from the approximate 6,000°C of the photosphere to at least 20,000°C.

The Discovery of the Chromosphere

The solar physicist Kevin Reardon of the U.S. National Solar Observatory has shown that people have commented on the presence of the chromosphere at the beginning and end of eclipse totality for over three hundred years. The Astronomer Royal John Flamsteed reported to the Royal Society in 1706 that a Captain Stanyan reported that 'his getting out of his eclipse was preceded by a blood-red streak from its left limb which continued not longer than six or seven seconds of time; then part of the Sun's disc appeared all of a sudden'. But saying that they saw this reddishness didn't mean that they understood it: Stanyan continued, 'I take notice of it to you, because it infers that the Moon has an atmosphere.' Other observers made similar comments in the early eighteenth century, including Edmond Halley, who made special preparations to observe the 1715 total solar eclipse that crossed England.

Over a century later, another Astronomer Royal, George Biddell Airy, also saw reddishness around the edge of the sun at an eclipse. When such a red feature sticks up substantially, it is known as a 'prominence' (illus. 40). (When someone casually refers to such a thing as a 'flare' on the edge of the sun, they are mistaken; solar flares are entirely different things.) In the *Memoirs of the Royal Astronomical Society*, Airy wrote, in 1842,

> While thus looking at the moon I saw, to my great surprise, some small red flames at the apparent bottom of the disc . . . their colour was a full lake-red; and their brilliancy greater than that of any other part of the ring.

His account seems to imply that he saw the reddish chromosphere as well. At an eclipse, the chromosphere is about 1/1,000th as bright as the solar photosphere. In turn, the corona is another factor of about 1,000 fainter than the chromosphere.

The solar physicists Peter Foukal and Jack Eddy have commented that reports of the solar chromosphere during the Maunder Minimum (1645–1715) of the sunspot cycle – the extended time period when the number of sunspots was extremely low – indicates that some magnetic activity persisted on the Sun even during that phase, given that the chromospheric structure is held in place and perhaps created by magnetic fields on the Sun.

40 The solar chromosphere and prominences, from Father Angelo Secchi's 1875 book *Le Soleil* (and its German version, *Die Sonne*).

Details of the Chromosphere

At the 1851 eclipse, Airy and the English astronomer Francis Baily commented on not only the prominences but 'a rugged line of projections . . . more brilliant than the other prominences . . . its colour was nearly scarlet'. They spoke of this saw-toothed appearance as a 'sierra' (illus. 41).

By the time of the 1860 eclipse, photography was able to show the chromosphere (though still only in black-and white; it was decades before colour photography). Comparison of photographs by the Englishman Warren De La Rue and the Italian Jesuit Father Angelo Secchi showed that the prominences moved with the Sun, not the Moon, and so were solar and not part of any lunar atmosphere. (We now know that the Moon does not have an atmosphere.)

At the 1860 eclipse in Spain, Secchi even commented that 'this material covers all the solar surface, like an overall transparent envelope. It is then beyond doubt that the Sun is enveloped at the limit of the photosphere by a sort of wrapping of a weak light of pink transparent gas, that is otherwise invisible in all normal observations.'

41 The solar chromosphere as a 'burning prairie', an 1872 diagram from the book *The Sun* by the American astronomer C. A. Young (1881).

Spectroscopy and the Discovery of Helium

The 1868 total solar eclipse that crossed southern Asia was a turning point in the history of our knowledge of the chromosphere. Several expeditions went to India from England and France for the event. The best known was that of Jules Janssen, the French astronomer. He took the recently developed spectroscope to an eclipse for the first time. During the eclipse, when the Fraunhofer (absorption) spectrum of the solar photosphere disappeared as the Moon completely covered the solar photosphere, a series of coloured lines appeared across the spectrum. These are called emission lines, since they appear against a black (or at least relatively dark) spectrum and represent the emission of radiation at those specific colours (wavelengths). Among the most prominent emission lines were a pair of bright yellow (or orange, some say) lines that corresponded to Fraunhofer's D absorption lines; those lines were thought to be from sodium (the so-called 'sodium-D lines') and explain why a flame glows yellow when you drop salt (sodium chloride = NaCl) into it. Janssen realized that the yellow line he saw in the spectrum of the chromosphere was not quite at the wavelength (colour) of the sodium-D lines, but was slightly displaced. Just as significantly, Janssen realized that the emission lines appeared so bright that an eclipse might not be required for one to see them, and on a subsequent day he managed to do so.

But the history is more complicated than the widely known version of the event, and Janssen was not the only one to see this bright yellow line at the eclipse. The English astronomers Norman Pogson, James Tennant and John Herschel were also using spectroscopes in India, and also saw this spectroscopic emission. The 2013 book by Biman Nath, *The Story of Helium and the Birth of Astrophysics*, tries to set the story straight.

At the time of this eclipse, the English astronomer Norman Lockyer was recuperating from an illness and didn't go to India for the event. But he had ordered a more powerful spectroscope than had been taken to the eclipse. It was not until months later that Lockyer received the spectroscope he had ordered, but then he was able to see the chromospheric and prominence spectrum from his site in England without the need for an eclipse.

Lockyer worked in the laboratory with the eminent chemist Edward Frankland, but the yellow line did not seem to match any known spectral line. So they said that the spectral line was from 'helium', since it existed only on the Sun (using the name of Helios, the Greek sun god). Since the terrestrial sodium-D lines could be called D1 and D2, this new line is known as D3 (illus. 42).

Lockyer was not the nicest person, leading the eminent Scottish astronomer James Clerk Maxwell (usually considered one of the

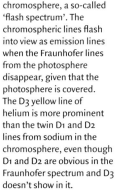

42 A modern eclipse spectrum of the solar chromosphere, a so-called 'flash spectrum'. The chromospheric lines flash into view as emission lines when the Fraunhofer lines from the photosphere disappear, given that the photosphere is covered. The D3 yellow line of helium is more prominent than the twin D1 and D2 lines from sodium in the chromosphere, even though D1 and D2 are obvious in the Fraunhofer spectrum and D3 doesn't show in it.

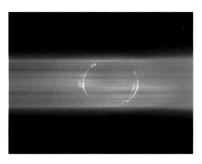
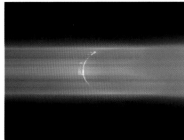
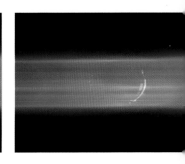

43 Medal of the mint in Paris commemorating the discoveries of Janssen and Lockyer. The obverse shows Janssen and Lockyer while the reverse of the coin shows the prominences coming from the eclipsed Sun.

greatest physicists of all time, along with Newton and Einstein) to write:

> And Lockyer, and Lockyer
> Gets cockier, and cockier
> For he thinks he's the owner
> Of the solar corona.

Though Lockyer's observations, on 19 October 1868, were made two months and a day after Janssen took advantage of the emission lines' brightness to see them the day after the eclipse, coincidentally the papers they submitted to the Académie des Sciences in Paris arrived on the same day. So it is generally said that the discoveries of Janssen and Lockyer were simultaneous (omitting mention of the other eclipse observers of the yellow line). Lockyer wrote (in a paper in 1868 in the *Proceedings of the Royal Society of London*) that

> these observations include the discovery, and exact
> determination of the lines, of the prominence-spectrum
> on the 20th of October, and of the fact that the prominences

are merely local aggregations of a gaseous medium which entirely envelopes the sun. The term *Chromosphere* is suggested for this envelope, in order to distinguish it from the cool absorbing atmosphere on the one hand [presumably the source of the Fraunhofer lines, which is now realized to be the upper part of the photosphere], and from the white light-giving photosphere on the other.

It was not until 1895 that the British chemist William Ramsay was able to isolate the gas helium on Earth; we now know that it is the second simplest element, with two protons and two neutrons surrounded by two electrons in its neutral state. It is a transition in the neutral state that leads to the so-called D3 helium line; the energy levels of helium are such that it takes a higher temperature than that of the photosphere to raise electrons to the energy level that leads to D3 emission.

Viewing the Chromosphere and Prominences

The solar chromosphere is so thin in its continuous radiation that we see right through it, down to the solar photosphere. Or, to put it the other way around, the solar photospheric light comes through the chromosphere with minimal disturbance en route to us on Earth. But if we look at the chromosphere with a filter at the wavelength of one of the strongest chromospheric lines, the absorption adds up and our ability to see in stops at the level of the chromosphere. The reddish colour of the photosphere is due to the strength of the emission from hydrogen at a red wavelength, known as Hα because it is the first of a particular series of emission lines coming from atomic transitions of hydrogen's single electron. Therefore, if you take an image of the chromosphere at the red wavelength of Hα, what you see is the chromosphere. Similarly for an image at one of the ultraviolet wavelengths of ionized calcium, you can view the Fraunhofer

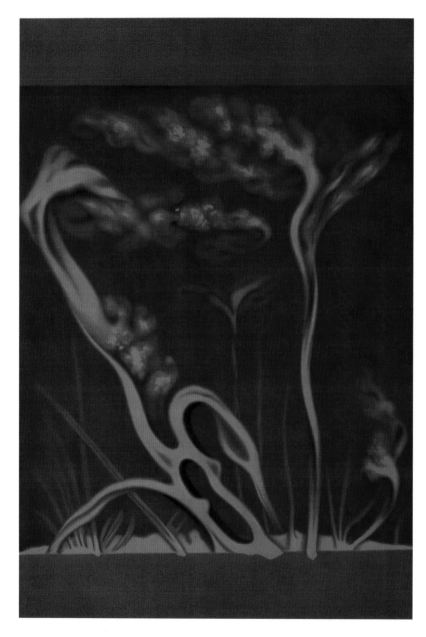

44 Lithograph showing a drawing by Étienne-Léopold Trouvelot of solar prominences, seen in Hα, 1881.

K line, a spectral line in the violet emitted by atoms of calcium and formed at a higher temperature, corresponding to a greater height in the chromosphere. (The Ca line labelled H by Fraunhofer is a bit longer in wavelength and thus somewhat easier to see, but it is blended with an unrelated line of hydrogen close to its central wavelength.)

Gianna Cauzzi of the Arcetri Astrophysical Observatory in Florence, Italy (and now of the National Solar Observatory in Boulder), and Reardon, with Dutch and Norwegian colleagues, have used an imaging spectrometer at the Dunn Solar Telescope at the Sacramento Peak Observatory in New Mexico to compare high spatial and high temporal observations of the hydrogen and calcium images to assess mechanisms of chromospheric heating.

The Twentieth-century Chromosphere

The mid-twentieth century brought great observational advances to solar physics, including the exquisite seeing (basically, atmospheric steadiness) at such sites (though only for limited times of day) as

45 *Left*: An image of the Sun through a hydrogen-alpha filter shows structure in the solar chromosphere. In addition to the background chromosphere we see dark filaments, sunspots and bright ribbons in the active region at lower right, indicating that a large flare is in progress. *Right*: A near simultaneous image of the Sun through a calcium K-line filter shows structure in the solar chromosphere. We see bright regions known as plages (pronounced 'plaahj', French for beach) in addition to the background chromosphere. At this chromospheric height, we see supergranules outlined in the plages surrounding sunspots. The flare ribbon brightenings are also visible, though less dramatically than in Hα.

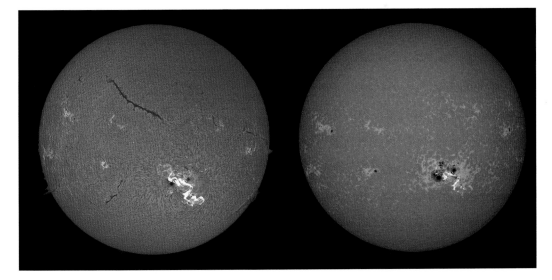

the Sacramento Peak Observatory in New Mexico. Richard Dunn
built instruments there that could see the chromosphere at higher
resolution than previously, as did Rawi Bhavilai from Thailand
and Jacques Beckers from the Netherlands while both worked in
Australia, building on work by Walter Roberts at the High Altitude
Observatory in Climax, Colorado.

These observers could not only see but could record photo-
graphically that the chromosphere is not a uniform layer of the
solar atmosphere (even if theoreticians often treated it as such)
but is broken up into tiny spikes, which Roberts named 'spicules'.
These spicules rise and fall in about fifteen minutes, and there are
hundreds of thousands of them on the Sun at all times. In 1965
Robert Noyes of the Harvard College Observatory and one of us
(JMP, for his PhD thesis) joined Beckers at Sacramento Peak to
measure spicule qualities, such as their height and width, which
was at the limit of the resolution. The spicules seen at the solar
limb, because they were at the limit of atmospheric 'seeing', were
almost always burred with neighbouring spicules. To make it worse,
they would be seen in projection at the edge of the Sun, always with
foreground and background spicules that weren't quite on the limb.
Studying individual spicules was not possible. It could be seen that
they are about ten times taller, at their maximum altitude, than they
are wide.

The amount of mass in the spicules is enough to replenish
the solar corona in less than an hour, so, clearly, much of the
mass falls back down. But it is extremely difficult to follow the
evolution of the spicules by direct observation. A large fraction
of the spicules are seen to disappear from the top down, a visual
effect that could result from the spicules disappearing as heating
from the top causes the spectral lines in which they are studied
– usually H-alpha – to disappear. Therefore, the Sacramento Peak
observations, and subsequent observations by JMP and others,
used spectroscopic methods to measure the velocities of the

46 Spicules near the solar north pole (the original location of spicules described by Walter Orr Roberts in 1945), observed with the Solar Optical Telescope on *Hinode* as a negative to bring out faint features and with special high-contrast processing using a method called Madmax developed by Serge Koutchmy of l'Institut d'Astrophysique, Paris. The resolution is on the order of 70 km or one-tenth of an arc second, about ten times better than traditional ground-based resolution. The whole frame is only 10 arcseconds wide, while the diameter of the solar disc is about 1,900 arcseconds.

spicules directly via Doppler shifts, though we could see only the components towards or away from us, while many spicules are obviously slanted. Zadig Mouradian at Meudon, France, also observed spicules spectroscopically; Serge Koutchmy in Paris has used computer algorithms to make high-contrast, high-resolution images of spicules (illus. 46).

Beckers's summary of the properties of spicules that appeared in the journal *Solar Physics* in 1968 became the bible for future studies of spicules. Only in the 1990s did Alphonse Sterling of NASA's Goddard Space Flight Center take up the study of spicules and their properties. JMP and his students further modernized spicule statistics with a series of NASA-sponsored observations with the Swedish 1-m Solar Telescope on La Palma in the Canary Islands in the 2000s.

Over many of those years, the designs of NASA spacecraft concentrated on expanding the spectral range under study, including especially x-ray and ultraviolet imaging, but at the cost of spatial resolution. A proposal to NASA for a high-resolution Solar Optical Telescope did not succeed. Only in the 2000s, did NASA's Transition Region and Coronal Explorer (TRACE) and, later, NASA's Solar Dynamics Observatory, along with the Solar Optical Telescope (illus. 47), on Japan's *Hinode* spacecraft bring high spatial resolution to space, reaching the limit that had long been barely obtainable from Earth, though with steadier images, given that the vagaries of the terrestrial atmosphere were removed. We will discuss these important solar spacecraft in a later chapter.

The spacecraft above the Earth's atmosphere, by being sensitive in the extreme ultraviolet, can observe a more fundamental emission line of helium from the first transition of its orbital electron up from the ground state, which appears at only

304 Å, about one-twentieth the wavelength of Hα of hydrogen. It can also observe the equivalent, more fundamental line of hydrogen than Hα, the so-called Lyman-alpha emission line at about one-fifth the wavelength of Hα. Important as these lines are, radiation at neither of these ultraviolet (UV) and extreme ultraviolet (EUV) wavelengths comes through the Earth's atmosphere, so we need to put our instruments in space if we want to detect this radiation.

One of the limitations of ground-based observing was the variability of imaging, with occasional bits of excellent seeing mixed in moment by moment at unpredictable times. A key to improving the situation had been discovered at Sacramento Peak, supposedly when it was noticed that the seeing improved when the lawn was being watered. It was realized that a body of water provides a less disturbed laminar flow of air and eliminates rising currents of hot air that cause turbulence in the view of the telescope. This discovery led to the erection of solar telescopes alongside or even in a body of water. Harold Zirin of Caltech built the Big Bear Solar Observatory at the end of a causeway out into Big Bear Lake in California to provide steady images over a big fraction of the day,

47 Image from the Solar Optical Telescope on *Hinode*, 2013, showing a prominence (at some angles, prominences appear as 'filaments', the term used when one is seen in projection against the solar disc) above a forest of spicules, imaged in the ultraviolet.

allowing him and his students and colleagues to make films in Hα of chromospheric activity. These often showed the everyday motions of spicules and other phenomena that are known as the quiet Sun, but also included violent activity: solar flares. At that time, films were made with the same kind of cameras used in Hollywood, and were recorded on film that had to be processed and studied on its reels. The long-term Chief Observer at Big Bear, Arvind Bhatnagar, built a similar observatory on an island in a lake in Udaipur, India.

The Big Bear Solar Observatory was taken over by the New Jersey Institute of Technology on Zirin's retirement, in the hands of Haimin Wang, who had worked with Zirin. It now boasts a 1.6-m New Solar Telescope installed in 2010, with a sophisticated reimaging system to discard most of the solar heat, and adaptive optics.

In 2007 Bart De Pontieu of the Lockheed Solar and Astrophysics Laboratory, which runs several space solar telescopes, reported that in addition to the regular 'Type I' spicule, with an average lifetime of fifteen minutes, which had been studied since their naming by Roberts in 1945, there is a 'Type II' spicule that quickly fades into

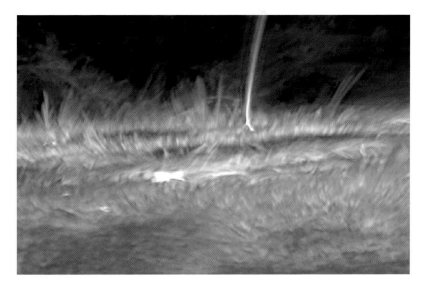

48 Spicules at the solar limb, imaged with the Solar Optical Telescope on *Hinode*.

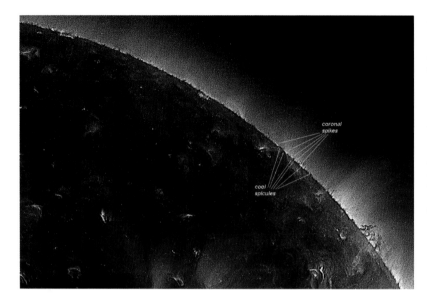

49 In this highly processed image from the Solar Dynamics Observatory's Atmospheric Imaging Assembly at a wavelength of 193 Å, which includes largely radiation from 11-times ionized iron (emitted from a gas at a temperature of 1.2 million K), it appears that the bright points that represent the bottoms of plumes of coronal gas are not quite at the same places as the dark points in the chromosphere that may represent the chromospheric spicules.

the corona after a shorter lifetime, only a few seconds to about two minutes. These features are best studied from space with the Solar Optical Telescope on *Hinode* because of its spatial resolution, which allowed De Pontieu and collaborators to see spicules three or four times narrower than had been previously observed, only about 200 km wide, and moving at far higher speeds than the others. They interpret the spicules in terms of a type of magnetic wave passing through the chromosphere; the waves are called Alfvénic after Hannes Alfvén, recipient of the 1970 Nobel Prize in Physics for related studies. This type of wave is one of the competing mechanisms for heating the solar corona to its observed million-degree temperature.

Because the corona is so much hotter than the solar surface we know that some mechanism must feed energy into the corona from below. A crucial question in trying to understand how energy flows from the photosphere to heat the corona is the shape of the interface between the chromosphere and corona, the so-called transition region, because the competing theories for how the energy is transferred involve different geometries for the structures that are being heated (illus. 49). Astronomers can study the transition region by looking at images of ions that show temperatures hotter than those of the chromosphere but cooler than those of the corona, such as five-times-ionized oxygen. If the chromosphere is made entirely of spicules, with no inter-spicular medium, then the corona reaches down to the feet of the spicules, and the transition region is on the sides of the spicules. This question is actively being studied by a small *Explorer* satellite known as IRIS (Interface Region Imaging Spectrograph) launched by NASA in 2013. IRIS has both a spectrograph and imaging, all in the ultraviolet where the chromospheric and transition region spectral lines can be observed.

Can We See Spicules on the Solar Disc?

There has long been a controversy as to what features on the solar disc match the spicules that are seen on the solar limb. It is surprisingly difficult to answer this question, in large measure because any one spicule does not live long enough for us to see it make the transition from disc to limb. Trying to trace the bottoms of spicules onto the disc has given ambiguous results, sometimes indicating that the spicules correspond to small bright elements on the disc and sometimes to small dark ones. There are rosettes of dark features on the disc that may well be the spicules.

The Solar Telescope of the Future

The biggest project in solar physics is now the Daniel K. Inouye Solar Telescope (DKIST), renamed after the Hawaii senator after long being called the Advanced Technology Solar Telescope. Its dome already stands on the 3,000-m (10,000-ft) summit of the Haleakala dormant volcano on the Hawaiian island of Maui, and its telescope and other internal parts are under construction, with

completion expected in 2019. It is funded by the U.S. National
Science Foundation (illus. 50).

DKIST (properly pronounced with the letters sounded out:
D-K-I-S-T) will be able to image the solar chromosphere with
unprecedented resolution. Among its qualities is that it will have
the biggest mirror of any solar telescope, over four metres across.
It was long held that solar telescopes did not have to be as big
as night-time telescopes because the Sun is so bright, but solar
physicists now spread out the solar spectrum so much and make
the image size so big that they have become light-starved, especially
given the realization that exposures have to be short because the
spicules and other chromospheric elements change so quickly. In
order to deal with the blurring caused by our murky atmosphere,

50 The construction on
Haleakala, Maui, Hawaii,
of the Daniel K. Inouye
Solar Telescope (DKIST),
as of January 2016.

DKIST will have 'adaptive optics' – that is, its main mirror is thin enough, only 75 mm thick – that actuators behind it can deform it to compensate for irregularities produced in the incoming wavefronts by the overlying atmosphere. These techniques of adaptive optics also required the advances in rapid computing that have come in recent years. DKIST is supposed to resolve solar elements to about one-tenth of an arc second, about 70 km on the Sun, which is about ten times better than the old limit of ground-based excellent seeing.

51 Eclipse image, highly processed from a variety of images to bring out fine features in the corona.

SIX

THE VISIBLE CORONA

When we stand in the location of a total solar eclipse, we are in an ordinary-looking place and may have travelled halfway around the world. But we know – and trust – the scientists who have predicted that, at the anointed time, we will be in the shadow of the Moon. Though the Moon begins partly to cover the Sun an hour or more before totality, one wouldn't notice for a long time that anything is happening. People around you may be going about their business, completely unaware of the treat that lies ahead. But about fifteen minutes or so before totality, the light takes on an eerie quality, something you can't put your finger on. Only with hindsight might you realize that shadows have changed in some way: they are no longer being projected by the full disc of the Sun, a half-degree across, but by a narrow crescent, so the shadows look sharper.

For the last few minutes before totality, the sky grows noticeably darker. But the Sun is about a million times brighter than the full Moon, so even when only 1 per cent – one one-hundredth – of the everyday Sun is left, the residual is still 10,000 or so times what the full Moon's brightness would be at night, and it is still light outside. The tiny crescent of the Sun is small but still at full brightness, so it is still unsafe to stare at the residual solar disc directly; you need the increasingly available 'eclipse glasses' (which would be better described as 'partial-eclipse glasses' – see Appendices I and II for

guidelines to safe viewing) to see the crescent. Even a photograph would just image an overexposed blur without showing the solar crescent.

Then, things begin to change more quickly. The Moon almost entirely covers the Sun, with only a few beads of sunlight visible as the last bits of the everyday Sun are visible in the valleys aligned your way at the edge of the Moon. These 'beads' are named after the English astronomer Francis Baily, who saw and wrote about them at the 1836 eclipse. The beads had actually been seen and commented upon earlier by Harvard's Francis Williams at the 1780 eclipse seen from what we now call Maine, though it was then part of Massachusetts and behind enemy (that is, British) lines for the American astronomers.

The last bead of sunlight – and there is usually only one – is so beautiful and relatively bright compared to the darkening sky that it has been known, since the 1925 total solar eclipse in New York City, as

52 During totality in a total solar eclipse, the solar corona surrounds the silhouette of the Moon, which covers an angle of only about one-quarter the diameter of your thumb at the end of your outstretched arm.

the diamond-ring effect. Only at this time can you safely take off your partial-eclipse glasses and look at the celestial phenomenon directly.

As the diamond ring disappears, a halo becomes visible around the Sun (illus. 52). It is known as the corona, after the Latin word for 'crown'. Edmond Halley, better known for his work on comets, described it as 'pearly white', and that description has stuck.

Discovering the Corona

Solar eclipses have been known for thousands of years. Indeed, the projection of a non-round, crescent-shaped image during an eclipse using what we call a pinhole camera may have helped lead to the discovery of the laws of optics and imaging. (The interstices between the leaves on a tree or a tent-peg hole may have acted as the hole that made the projection; the term 'pinhole' probably falsely implies a smaller hole than is used or usable.) So accidental imaging during an eclipse probably started the series of steps in optical imaging that eventually led to the still camera and eventually the film camera and video camera, bringing us up to the cinema and television of today – distant descendants of solar eclipse phenomena.

But for thousands of years, people commented on or measured only the timing of an eclipse, and printed illustrations showed only solar crescents (illus. 53). The extent of the crescent and whether or not the Sun is completely covered at a given location is something that can be written down and preserved through the ages. F. R. Stephenson of Durham University has been able to use mentions of solar eclipses from over two thousand years ago to calculate changes in the Earth's rotation rate over time. After all, the Earth rotates around 30 km each minute at the equator, and it is quite noticeable whether or not the Sun is totally eclipsed at a given location, so even general comments in an ancient manuscript can reveal whether or not there was totality. We can calculate very accurately what rotation rate of the Earth would or would not have produced totality at a given location.

Eclipsis solis
dies ho m z? pu m ho m

Julius	1478	29 1 48	0	8 45	1 52		
Julius	1479	18 17 26	0	0 38	0 38		
December	1479	12 23 40	24	7 41	2 6		
Maius	1481	28 6 20	13	2 3	1 18		
Maius	1482	17 7 41	41	5 0	1 50		
Marcius	1485	10 4 34	52	12 23	2 0		
Marcius	1486	5 17 47	0	9 21	2 0		
Julius	1487	20 2 9	0	7 5	1 44		
Julius	1488	8 17 36	0	3 11	1 18		
Maius	1491	8 3 18	0	8 55	2 10		
October	1492	20 23 14	18	2 12	1 16		

Eclipsis solis
dies ho m z? pu m ho m

October	1493	10 2 35	0	8 26	2 10		
Marcius	1494	7 6 10	0	2 17	1 8		
Julius	1497	29 2 58	0	3 38	1 22		
September	1502	30 19 30	0	9 8	2 14		
Julius	1506	20 3 1	0	2 33	1 20		
Marcius	1513	7 1 41	0	5 31	1 42		
December	1516	23 3 48	22	3 10	1 24		
Junius	1518	7 17 46	1	10 34	2 12		
October	1519	23 4 30	22	6 10	1 58		
October	1520	11 5 23	17	3 16	1 36		
Augustus	1532	30 0 59	0	5 42	1 40		

53 Page of solar-eclipse images from the fifteenth century: a *Calendarium* by Pflaum.

The first clear mentions of the solar corona were apparently made by the great astronomer Johannes Kepler in 1604, in his *Astronomiae pars optica* (The Optical Part of Astronomy), and in 1606, in his book describing the supernova of 1604. In 1605, Kepler had written a sixteen-page pamphlet about eclipses, describing a series of eclipses, including the one of that year. But that pamphlet does not describe anything about the appearance of the eclipse. On the other hand, in his 1606 book about the 'new star' Kepler wrote, here translated from the Latin (illus. 54):

The whole Sun was effectively covered, but indeed it did not last for a long time. In the middle, where the Moon was, there was the appearance as if of a black cloud; all around there was a reddish and flaming brilliance, of uniform breadth, which occupied a good part of the sky (trans. Edan Dekel, Williams College).

116 JOANNIS KEPPLERI

ca corpus Lunæ accensum suftinuit, materia cælestis fuit. Neapolitana verò relatio superioris anni sic habet : Accuratè rectum fuisse totum Solem , quod quidem non diu duraverit ; in medio , ubi Luna , fuisse speciem quasi nigræ nubis ; circumcirca rubentem & flammeum splendorem, æqualis undique latitudinis, qui bonam cœli partem occupaverit : E regione Solis, versus Septentrionem , cœlum obscurum planè, ut cùm profunda nox est; stellas tamen non visas. Ut autem nihil dubites de fide historiæ, ecce aliam ex Flandria; ubi non totus quidem Sol tectus; prominebat enim suprema pars circuli solaris lucida, latitudine unius digiti, aut dimidij (sanè quia Antverpiæ, citeriori loco , extabat digitus) : sed tamen globus Lunæ visus, declinans ad nigredinem, fuscus, aut quasi fuligine tectus; cùm superior circumferentia Lunæ esset tota candida, & quasi ignea. Et ut constaret visum esse locum disci Lunæ integrè circumscriptum; addit relatio, locum omnem, in quem à Sole visus aversus dirigeretur, visum esse fuliginosum, circumferentiâ igneâ. Non poterat igitur phænomenon ipsum habere aliter, cujus species in oculo talis erat. Simile quippiam visum est Jessenio Torgæ in Eclipsi anni 1598. Vidit enim splendore Lunam planè cingi. Vide pag. 299. Opticorum: ubi ultimam vocem aëris latè accipe pro ætherea etiam substantia.

Hic quæro, quinam fuerit ille splendor igneus, circumdans Lunam, quæ ad visum erat Sole major; quia totum Solem absconderat? Imò quinam ille splendor, qui Lunam ab inferiore limbo, quo Solem hæc ad unīus digiti latitudinem excedebat, nihilominus amplectebatur? Splendor erat Solis, inquis. Verùm, at non hoc quæritur, sed quænam materia, quodnam subjectum, in quo inhæsit iste Solis splendor? Ipsa namque per se lux digressa à suo corpore cerni non potest, nisi in subjecto; quia nuspiam consistit, nuspiam impingitur, nisi in *Splendor ille non erat aëris noftri, neq; Neapoli.* opaco quodam subjecto. Si dicas, aërem fuisse hujuc splendori pro subjecto, diversorum locorum experientijs diversimodè refutabere. In Schemate præsenti sit αβ Sol , δζ Luna , δηζ Conus umbræ : globus Telluris ει. Igitur Neapoli, qui locus concipiatur in ε, totus Sol latuit. At ubi Sol latet, is locus in umbra est Lunæ, puta intra ι, ε : quare illa portio aëris, per quàm species ignei splendoris, Lunam proximè circumdans, in oculos observatoris est delapsa, illa inquam portio aëris erat in umbra Lunæ intra ν η. Sol igitur aërem illum, cui tribuitur splendor iste ab opponente, non illustravit. Dicet forsan adversarius, margines Lunæ Solisque adeò præcisè invicem applicatos, ut non benè discerneretur, an aliquid de Sole superesset, cùm reverà aliquid

54 Johannes Kepler described the solar corona for the first time; here is a mention from *De Stella nova in pede serpentarii* (1606).

Aside from the reddish colour ascribed, and the rough size of the phenomenon, on the whole the description seems to match the solar corona and the various optical effects that precede and follow totality. But Kepler seems to think that we are seeing a lunar atmosphere being backlit by the bright, covered-up disc of the Sun.

Wherefore the Corona?

There may be a halo, a corona, around the lunar silhouette during a total solar eclipse, but what is it exactly? Is it the atmosphere of the Moon? Even in the twenty-first century, some people (not astronomers!) mistakenly believe so. The eclipses of 1715 (illus. 55) and 1724 were observed from widely separated locations in Europe, and the appearance of the corona was the same at both locations. Had the corona been located on the Moon, only 400,000 km away from us on Earth, instead of on the Sun, 150,000,000 km away (about 400 times farther), the corona would have appeared shifted over as viewed from one site compared to the other, an effect known as parallax. (See the parallax effect by looking at a distant object behind your thumb at the end of your outstretched arm, though this time first from one eye and then from the other.) Also, the Moon was seen to move across the corona, rather than having the corona move with the Moon. So the corona was apparently associated with the Sun itself rather than being part of the Moon. This point remained controversial for decades. The photographic proof at the 1860 eclipse that bright solar prominences at the edge of the lunar silhouette did not show parallax helped convince scientists that the corona was solar.

What is the Corona Made Of?

In the nineteenth century, expeditions were mounted that travelled to study eclipses. The 1868 eclipse is particularly well known for

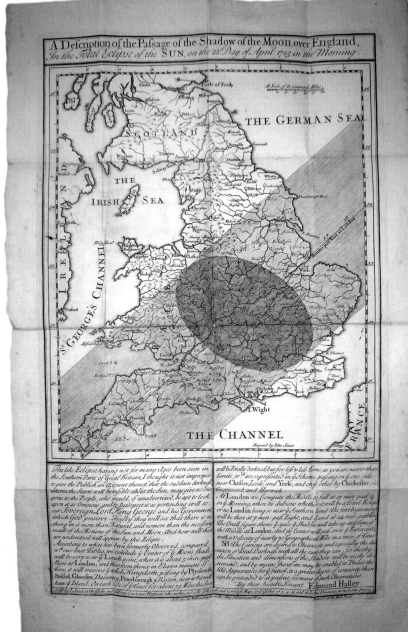

55 Edmond Halley's map of the path of totality at the 1715 eclipse was the first such depiction of an eclipse path. After the eclipse, he published a slight correction to the eclipse path as actually observed and provided a prediction of the 1724 path, also across Europe.

the several expeditions that went to India to study it. (It also went through Siam, now Thailand; the famous King Mongkut of Siam, as immortalized in the Rodgers and Hammerstein musical *The King and I*, actually died of malaria contracted during his trip to the path of totality through his country.)

At the 1868 eclipse, the newly developed spectroscope was taken to India, and a strange yellow spectral line appeared in the spectrum at the edge of the Moon. It wasn't quite at the position of the expected yellow pair of lines from sodium, Fraunhofer's D line, and implied the presence of a new element named helium, as we discussed in the last chapter. It wasn't until 1895 that chemists isolated helium on Earth. Of course, we now know that it comes right after hydrogen on the periodic table of the elements. At the following eclipse, in 1869 in the United States, spectroscopists saw a green spectral line in the solar corona (illus. 56). By analogy to the name helium, it became called 'coronium', since it apparently occurred only in the corona. But whereas helium found its place in the periodic table, the table became filled in and there was no room for coronium. It was another seventy years before the mystery of coronium was solved.

Overall, it wasn't appreciated at the time that the Sun and the other stars are almost entirely hydrogen, as we now say: 90 per cent hydrogen, 9 per cent helium, and less than 1 per cent everything else. For example, the observation of many spectral lines of iron led to misleading conclusions about a relatively high abundance of iron in the Sun, suggesting a much higher value than we now know is the case. Only with the Radcliffe College PhD thesis of Cecilia Payne (later known as Cecilia Payne-Gaposchkin) was an improved type of calculation done and the conclusion drawn that the Sun and stars are largely made up of hydrogen. Even then, Payne's conclusion was doubted by the influential Princeton astronomer Henry Norris Russell. Only after a couple of years, with new calculations by Donald H. Menzel, did Russell come around to agree with Payne's now universally accepted conclusion. (Menzel came to Harvard at

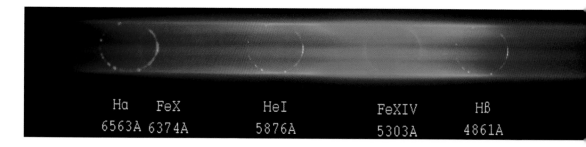

Hα	FeX	HeI	FeXIV	Hβ
6563A	6374A	5876A	5303A	4861A

56 Spectrum of the corona at the 2015 total solar eclipse showing the green emission line at 530.3 nm, once known as coronium.

about the time of the 1932 total solar eclipse and wound up, in the 1950s, as Director, championing Payne-Gaposchkin and seeing that she was given her deserved rank as Professor of Astronomy. In 1976 she was awarded the honour of giving the Henry Norris Russell Lecture of the American Astronomical Society.)

Spectroscopy of the corona is particularly fruitful in the infrared. There is a triplet of infrared lines from Fe XIII (twelve-times-ionized iron; recall that an un-ionized element has a spectrum designated as I, for example, Fe I for the spectrum of neutral iron), and recent electronic advances with detectors have allowed imaging to wavelengths of several microns (a micron is the old name for a micrometre, which is a millionth of a metre), a few times longer than red light. One of us (LG) will observe the infrared spectrum in this region from an instrumented aeroplane during the 2017 total solar eclipse.

Most of the coronal lines known in the visible spectrum are from highly ionized iron or calcium, with a few lines from silicon or sulphur. But all these lines are subsidiary; the primary spectral lines from the corona are in the extreme ultraviolet or x-ray regions of the spectrum, observable from spacecraft or sounding rockets, as we describe in the next chapter.

How Hot is the Corona?

The innermost corona shows only a few spectral lines and those appear as bright lines (emission lines) at eclipses, while the rest of the coronal spectrum looks very much like that of the 5,800 K photosphere in terms of its overall shape. It seems that the corona scatters the bright light of the solar disc towards us. But why don't we see the Fraunhofer absorption lines in the coronal spectrum?

To answer that question we need to understand what the term 'hot' means. To a physicist, it means that particles are moving very fast, to and fro, with their speed increasing as the temperature goes up. As the solar spectrum is scattered towards us, some of the coronal particles that are reflecting the light are moving towards us and others away from us. If the temperature is very high these motions are very fast, and the Doppler shift – changes in wavelength resulting from motion towards or away from the observer – spread out ('broaden') the absorption lines so much that they blend into the continuous background spectrum and can't be seen.

The credit for determining that the corona is millions of degrees (it hardly seems worthwhile to take account of the 273°C difference between the Kelvin scale that begins at absolute zero and the Celsius scale that begins at the freezing point of water) is usually assigned to Bengt Edlén in 1943. Edlén considered the coronal spectrum, which showed a handful of emission lines, chiefly one in the red and one in the green. Gas at millions of degrees is highly ionized, separated into protons, extremely ionized atoms and electrons. Such highly ionized gas is called 'plasma', though often you see plasma considered as a fourth state of matter, alongside solid, liquid and gas. No laboratory then could duplicate the spectrum of iron at millions of degrees, but Edlén worked out several series of lines along isoelectronic sequences: that is, sequences of different chemical elements, in order along the periodic table, with the same number of electrons, so that the successively higher atomic number elements are each one step more ionized than the previous one. Such sequences can be used to predict the unknown spectra if one knows other spectra in the sequence.

Walter Grotrian had shown in 1939 that the coronal red line is from Fe X, a state of iron having lost nine electrons. Edlén was able to determine that the coronal green line comes from iron that has lost thirteen of its electrons, half of the 26-electron complement of neutral iron. For that to be the case, the temperature has to be well over a million degrees. Edlén received the Gold Medal of the Royal Astronomical Society in 1945 for his work in helping to solve what had been known as the Corona Mystery.

In fact, neither Grotrian nor Edlén initially stated explicitly that the corona is at such high temperatures. In 1941, the Nobel-prize-winning physicist Hannes Alfvén, in a Swedish journal, examined the available evidence and summarized six arguments that the corona is 'heated to an extremely high temperature'. Alfvén is associated with championing the importance of magnetic fields and electric currents in astrophysical situations, and the magnetic field is indeed important in the heating of the corona. His conclusions on the temperature of the solar corona used to be routinely referenced, but are now not as widely known; his work was championed in 2014 in an article by H. Peter of Germany and B. Dwivedi of India.

Observing the Corona Outside of Eclipses

Though the longest history of observing the corona is in visible light, with the human eye or equivalent cameras, most of the radiation from the corona is at shorter wavelengths: in the ultraviolet or in x-rays. In this chapter we discuss the long history of observing the corona in visible light, and the next chapter will describe observing the corona in shorter wavelengths.

It isn't always appreciated, even today, that the corona is always up in the sky. It is merely much fainter than the blue sky, so it is only when the scattered light that produces the blue sky is taken away in the middle of the day – which is what happens when the Moon blocks ordinary sunlight from hitting air particles – that the faint corona becomes visible.

Bringing your instrument up on a high mountain with clean air is a help, but it isn't enough to allow you to see the corona outside of eclipse. Enough air remains to scatter the bright sunlight, limiting the visibility of the image in the instrument. But from space – as for astronauts on the Moon or on the International Space Station – you can block out the solar disc and see the corona.

In 1936, the talented instrument specialist Bernhard Lyot, in France, worked out a way to suppress the scattering of the bright photospheric light enough in a telescope to image the corona without an eclipse. Lyot used lenses rather than mirrors, which scattered light from minuscule irregularities in their coatings. He polished the lenses very carefully – notably using a bit of 'nose oil' (obtained by touching the side of his nose with a finger) spread out uniformly on the lens. Within the telescope, he carefully set up precisely positioned small absorbing components (generally known as 'Lyot stops') to block light reflected back and forth between the optical surfaces and diffracted around the edges of the optical elements. In this way, he could observe the inner corona from a high mountain. He also viewed a coronal emission line combined with a narrowband filter (still known today as a 'Lyot filter') centred at the wavelength of that line to provide an increase in relative brightness compared with the continuous solar radiation.

With Lyot's advance, the innermost corona could be monitored over the course of the sunspot cycle. Some of the locations with coronagraphs were the High Altitude Observatory's station at Climax, Colorado; the Sacramento Peak Observatory at Sunspot, New Mexico; and the Observatoire du Pic du Midi, France. The current best coronagraph site is on Mauna Loa, a nearly 4,000-m (14,000-ft) mountaintop on the Big Island of Hawaii (illus. 57). Even so, the best coronagraph images from such mountaintop observatories do not match the detail or extent of the corona as seen from eclipses. Nor do the images extend down quite to the surface of the everyday Sun; the bright photosphere needs to be covered up, so there is always a gap of some size, which can be filled in best at eclipses on those special eclipse days every eighteen months or so.

A few coronagraphs are in space, above the Earth's atmosphere. The most famous are aboard the Solar and Heliospheric Observatory (SOHO), a European Space Agency satellite launched in 1995 that carries coronagraphs built and operated by the U.S. Naval Research Laboratory. (As we will see in the next chapter, the NRL has been in the solar/space business since they were invited to use some of the German V-2 rockets for experimentation at the end of the Second World War.) It was built in collaboration with the University of

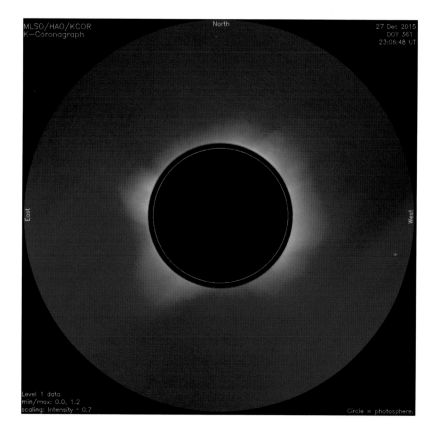

57 A current coronagraph image from the High Altitude Observatory's coronagraph on Mauna Loa, Hawaii. The white circle indicates where the edge of the photosphere is located behind the occulting disc.

Birmingham in the UK, the Laboratoire d'Astronomie Spatiale in France and the Max-Planck-Institut für Aeronomie in Germany.

SOHO carries three coronagraphs covering successively larger fields of view. From inside out in the solar corona, they are called C1, C2 and C3. C1 is the only one that was of the Lyot type, with an internal occulting disc to block the image of the solar photosphere. It suffered from too much internal scattering, even with the Lyot design, to be really useful. And then in 1989 the SOHO spacecraft spun out of control when its guidance system was confused by the non-round sun that was all that was visible during a solar eclipse. The spacecraft cooled, and when an astronaut managed to tame it and bring it back to life, the C1 coronagraph remained dead. (One of us arranged some eclipse observations with a camera built to match C1's field of view, to compare C1's scattered light with the coronal view taken with the dark background sky provided from Earth only during total eclipses.) The details of the NRL's Large Angle and Spectrometric Coronagraph Experiment (LASCO) are that C1 images from 1.1 to 3 solar radii (Max-Planck); C2 images from 1.5 to 6 solar radii (Astronomie Spatiale); and C3 images from 3.5 to 30 solar radii (NRL). Note that the innermost tenth of a solar radius above the solar limb remained exclusively in the eclipse capability; none of LASCO's coronagraphs covered that region.

SOHO is quite old for a spacecraft, and when it becomes defunct there will be no coronagraph in Earth orbit. Already the ultraviolet imaging on SOHO was transferred to an improved version on the Solar Dynamics Observatory.

In 2006, NASA launched STEREO, the Solar TErrestrial RElations Observatory, a pair of spacecraft each carrying a coronagraph. They are in orbit around the Sun, one slightly outside Earth's orbit and one slightly inside, so they have gradually moved around the Sun – one ahead of the Earth and one lagging behind the Earth – to give us views at different angles. As of 2016 they are each more than halfway around, and are not giving a very three-dimensional view. STEREO's Sun

58 A pair of Large Angle and
Spectrometric Coronagraph
Experiment (LASCO) views,
C2 at left and C3 at right,
from the SOHO spacecraft.

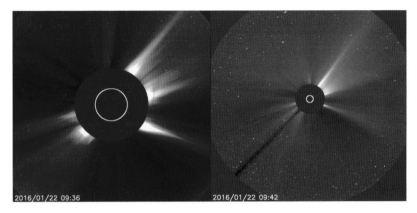

2016/01/22 09:36   2016/01/22 09:42

Earth Connection Coronal and Heliospheric Investigation (SECCHI)
has been given an acronym that matches the name of an important
nineteenth-century solar astronomer whose work we mentioned in
the previous chapter. It carries an inner coronagraph of Lyot type run
by NASA's Goddard Space Flight Center, an outer coronagraph run
by the Naval Research Laboratory and the Heliospheric Imager with
an even wider field of view. The inner coronagraph, COR1, has a field
of view from 1.3 to 4 solar radii. The outer coronagraph, COR2,
extends to 15 solar radii. The Helioseismic and Magnetic Imager
observes, between its two parts, out to Earth's orbit.

Imaging the Corona in Visible Light

The solar corona covers a wide range of brightness, falling in
brightness by a factor of about 1,000 from the edge of the Sun to
1 solar radius out, though the shape of the coronal features makes
that factor vary. Proceeding further out, the brightness continues to
fall so that no single imager can capture the whole range of coronal
brightness. Therefore even the details of coronal shapes are lost in
individual images.

The first coronal photograph was made in 1851, only a dozen
years after Louis-Jacques-Mandé Daguerre in France made his first

TOTAL ECLIPSE of the SUN.

Observed July 29, 1878, at Creston, Wyoming Territory.

photographs, working with François Arago of the Paris Observatory with the expectation of turning his new methods to astronomy. The 1851 coronal photo, a daguerreotype, is credited to 'Berkowitz' – probably Johann Julius Friedrich Berkowski, considered among the most skilled daguerreotypists in the city of Koenigsberg (now Kaliningrad).

Because the dynamic range of (formerly film and now) electronic detectors is much more limited than the corona's range of brightness, in recent years various computer techniques have been used to select the best parts of many images. But back in 1918, people

59 Lithograph of 1881 showing a stylized drawing by Étienne-Léopold Trouvelot of a total solar eclipse. The drawing corresponds to the view of the eclipse near the minimum of the sunspot cycle, with equatorial streamers and polar plumes. The reddish chromosphere and a prominence are also shown.

in charge of the u.s. Naval Observatory's eclipse expedition to Oregon invited an artist, Howard Russell Butler, to join them. He had a method of taking notes on details and colour, filling in an oil painting later. His resultant colour painting shows much more detail on coronal shapes than was available in photography at the time. He went on to paint solar eclipses of 1923, 1925 and 1932 (illus. 60). The full-size originals, up to 2 m (6+ feet) high, belong to the American Museum of Natural History in New York, where they hung at the entrance to the Hayden Planetarium for decades. Several half-size versions also exist, including at the Franklin Institute in Philadelphia, at Princeton, and – not displayed – at the Staten Island Museum and at the Buffalo Museum of Science.

These days the major computer composites of eclipses, showing details to great distances from the solar surface and enhancing the contrast of coronal features, are being made by the Czech computer scientist Miloslav Druckmüller. He uses several dozen individual images taken with varying exposures. The individual images, even though they usually come from ordinary Nikon or Canon cameras, are respectfully treated with the full range of procedures ordinarily given to scientific data, such as subtracting 'dark' frames and 'bias' frames to minimize the background noise and variable response across the field of the cameras' imaging sensor and electronic readout. His eclipse photography website (www.zam.fme.

60 Howard Russell Butler's triptych of the 1918, 1923 and 1925 total solar eclipses, painted based on his notes and sketches during the eclipses.

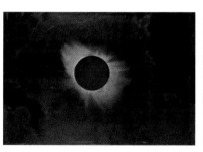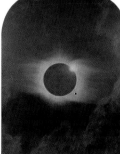

vutbr.cz/~druck/eclipse), easily findable with Google by typing in 'Druckmüller eclipse', includes eclipse images that he has reprocessed as far back as 1980.

Druckmüller now not only processes images taken according to a protocol of exposures that he prefers, such as some from one of us (JMP, illus. 51), but also travels to total solar eclipses himself, often accompanying the University of Hawaii solar scientist Shadia Habbal. Wendy Carlos also processes such multiple images (illus. 61).

The processed images show very clearly that a major part of the solar corona is the set of coronal streamers. Their shape is determined by the solar magnetic field and the way it interacts with the hot coronal gas/plasma. Their extent varies with the solar-activity cycle (most easily manifested to the eye as the sunspot cycle). At solar maximum, there are so many streamers pointing in all directions, like porcupine quills, that the overall corona looks fairly round; at solar minimum, there are mainly equatorial streamers, so the overall shape of the corona is substantially oval. At solar minimum, the absence of streamers at the poles reveals coronal plumes, narrow streams of gas held in place by the solar magnetic field. By comparing the changes in such plumes from different locations along the path of totality during a total eclipse, velocities of outflowing matter in the plumes can be measured.

There are windows of transparency in the Earth's atmosphere that include visible light, radio waves and parts of the infrared. No ultraviolet or x-rays come through our atmosphere to the ground, even to mountain top observatories. But since the 1940s, rockets have taken telescopes above the Earth's atmosphere to reveal the solar radiation in the short-wavelength part of the spectrum. This type of solar observation will be discussed in detail in the next chapter.

61 A composite image made of two dozen individual photographs by the New York musician and amateur astronomer Wendy Carlos, from photographs of the Zambia eclipse of 2001. We see a roundish, streamer-filled corona typical of solar maximum. Carlos's composites have less contrast and are closer in appearance to what is seen by eye than Druckmüller's, one of which was shown in illus. 51.

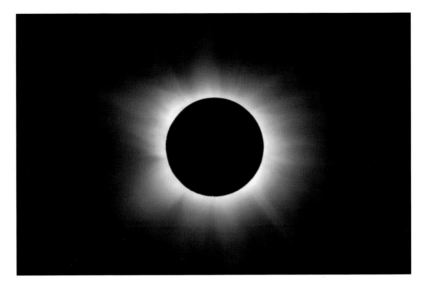

Imaging the Corona in Radio Waves

The hot coronal gas gives off radio waves at a wide variety of frequencies. The spatial details can be determined with arrays of radio telescopes, such as the Jansky Very Large Array in New Mexico, which one of us (JMP) used with colleagues to try to pinpoint the agreement between radio and x-ray points of emission on loops of coronal gas during the 2012 annular eclipse. Dedicated solar heliographs, composed of dozens of small radio telescopes electronically linked, exist in Japan and, most recently, at Mingantu in Inner Mongolia, China.

The major international radio-mapping telescopes known as the Atacama Large Millimeter/Submillimeter Array are equipped to study the Sun. We look forward to their solar mapping at the long-end of infrared radiation and the short-end of radio radiation, both at any scheduled times and also during the eclipses of the Sun that will be visible from their site in Chile in 2019 and 2020.

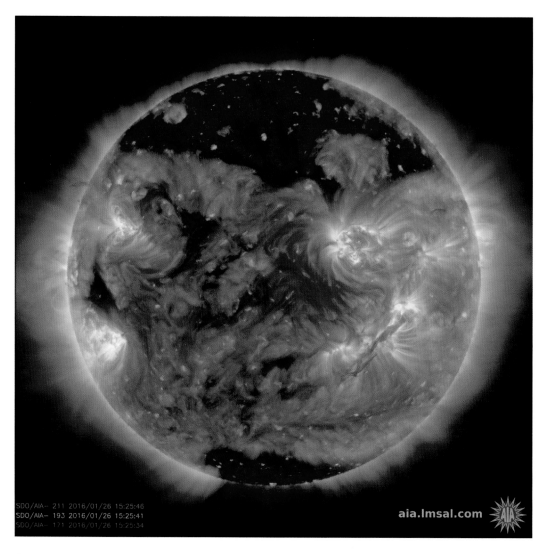

SDO/AIA– 211 2016/01/26 15:25:46
SDO/AIA– 193 2016/01/26 15:25:41
SDO/AIA– 171 2016/01/26 15:25:34

aia.lmsal.com

62 This image from the Atmospheric Imaging Assembly (AIA) on the Solar Dynamics Observatory shows the hot extreme ultraviolet (EUV) emission from the hot solar corona. Three images at different EUV wavelengths, corresponding to different temperatures, are combined.

SEVEN

THE INVISIBLE CORONA: A DISCUSSION MOSTLY ABOUT PHOTONS

What we call visible light is only a tiny fraction of the overall spectrum of electromagnetic wavelengths, which range from very long-wavelength radio waves at one end to extremely short-wavelength gamma rays at the other.[17] All of these are forms of electromagnetic waves emitted by charged particles when they are accelerated and which propagate through space at the speed of light. Our terrestrial atmosphere blocks many of the wavelengths that are either longer or shorter than the visible, so our perception of what is out in space was, until the advent of the Space Age, severely restricted. Once we attained the ability to put instruments above the atmosphere we became able to see these invisible wavelengths – invisible because they do not reach the ground and also because our perceptual systems did not evolve to detect colours beyond the red (infrared) and beyond the violet (ultraviolet). We saw a very different universe, filled with dynamic phenomena and often rapid and energetic changes. The old view of a slowly evolving, rather quiet universe changed to a realization that it is a dynamic and violent place, filled with exotic objects and hugely energetic outbursts. Nowhere has this change in our views been more clearly evident than on the Sun.

In 1879, about a decade after Maxwell's work that unified electrical and magnetic phenomena, a graduate student named

Heinrich Hertz had his supervisor, Hermann von Helmholtz, suggest that he work on the task of finding a way to test Maxwell's theory. Hertz did not immediately take up the challenge, but several years later he opted to determine whether any of the waves that were predicted by the theory could actually be detected. He built an apparatus that produced high-frequency radio waves and successfully detected them in a receiver placed several metres away. After carrying out a sophisticated series of tests on the properties of these waves (including a proof that they would reflect from a metallic surface) and publishing the results, Hertz seems to have lost interest in radio waves, reportedly telling some students who asked about the importance of his work: 'It's of no use whatsoever.'

Others, however, saw major applications. Through hard work, persistence and continual improvements in equipment, Guglielmo Giovanni Maria Marconi took command of the competitive field of wireless telegraphy. Starting in 1894 with experiments in his back garden and receiving a cool response in Italy, Marconi found a receptive audience for his technology in England and moved to London in 1896. He was able to transmit coded signals across the English Channel by 1899, and claimed a successful transatlantic signal reception in 1901, across a distance of 3,500 km. The claim is viewed with some scepticism today, since the broadcast took place in daytime, the worst possible time as it turns out, and at a bad choice of frequency as well. Marconi prepared a better test in 1902 using a receiver aboard ss *Philadelphia*, travelling west from the UK. The best transmissions were found to occur at night, reaching to 3,400 km, while during the day transmissions failed after about 1,100 km.

The problem with Marconi's startling result is that it should not have been possible, according to what was known at the time. Radio waves travel the same way that light does, mostly in a straight line if that path is not blocked in a given situation. Because of the curvature of the Earth, Marconi's signals should have been limited to at most 200 or 300 km, depending on the height of the antenna used to broadcast the signals. How could these waves be detected over the horizon to such great distances? A solution was offered nearly simultaneously on both sides of the Atlantic by an American electrical engineer, Arthur Kennelly, and by the British physicist Oliver Heaviside. They postulated a conducting layer high in the atmosphere that would reflect the radio waves, allowing them to ricochet around the curvature of the Earth. In 1912 the British physicist William Eccles proposed that the large difference between daytime and night-time transmission could be due to changes in the conducting layer caused by solar radiation. (T. S. Eliot – in a letter and later in a poem – referred to the 'Heaviside layer'; in the musical *Cats*, it is past the 'Jellicle moon', and therefore far away.)

The Kennelly–Heaviside layer, as it came to be called, was viewed with scepticism, and a theory involving diffraction of the radio waves was considered more likely. This theory predicted that long wavelengths, which diffract more easily than do short wavelengths, would be better suited to long-distance transmission. Government agencies then reserved those wavelengths and allowed the increasingly large numbers of amateur radio enthusiasts to operate in the 'useless' high-frequency, short-wave bands. It came as quite a shock therefore when, in November 1922, amateur operators achieved the first two-way communications across the Atlantic, between Nice, in France, and West Hartford, Connecticut. Because these transmissions were at short wavelengths, it was clear that they were not due to transmission along a path near the surface, but rather that they were bouncing from high in the atmosphere in what was called 'skip' transmission, in analogy with skipping a stone across the surface of a pond. Thus was interest in the atmospheric conducting layer renewed.

The Ionosphere

In Cambridge, UK, Edward Appleton decided to explore the
properties of the hypothetical reflecting layer by using a signal
broadcast from London by the BBC and examining the way that its
strength varied throughout the day and night. The signal received
at Cambridge consisted of both a direct beam and a reflected beam,
which interfered with each other because they travelled along different paths to arrive at the receiver (the same effect happens in visible
light, causing interference patterns, such as Newton's rings, to be
formed). On the night of 12 December 1924, he determined that the
layer, which he denoted 'E' for 'electric vector,' was at an altitude
of about 100 km. Continuing his experiments, he found another
layer above the E layer, and called it the F layer. Subsequently, he
found that another layer closer to the ground at about 60 km was
occasionally present; he chose to call this the D layer, rather than
renaming them 'A', 'B', 'C' and so forth, since he did not know
how many more layers he would find.

At about the same time in Washington, DC, Gregory Breit and
his student Merle Tuve, working with transmitting equipment
developed at the nearby Naval Research Laboratory (NRL), were
using a technique called 'pulse-height' to determine not only the
presence of the conducting layer, but its height. The key to this
method is that a short but intense pulse of radio waves is sent out
and the time needed to receive the reflected wave is measured. This
pulse method is the basis of the technique now known under the
acronym RADAR, for RAdio Detection And Ranging. Eventually
the Scottish physicist Robert Watson-Watt, who headed the UK's
radar development effort, suggested that the conducting layer be
called the ionosphere, in analogy with the terms 'stratosphere' and
'troposphere', and this became the accepted terminology (illus. 63).

The NRL had long been involved in the development of
military radio and radar equipment, and in studying the effects

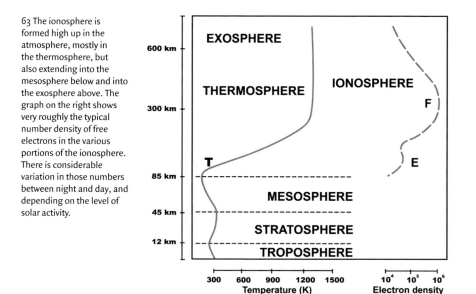

63 The ionosphere is formed high up in the atmosphere, mostly in the thermosphere, but also extending into the mesosphere below and into the exosphere above. The graph on the right shows very roughly the typical number density of free electrons in the various portions of the ionosphere. There is considerable variation in those numbers between night and day, and depending on the level of solar activity.

of the ionosphere on communications. As early as the 1920s, Edward Hulburt of the NRL had proposed that extreme ultraviolet (EUV) radiation from the Sun would produce an ionosphere layer as it is absorbed high in the atmosphere. It was thus natural that when, in 1945, Ernst Krause of NRL was invited to join a party being sent to Germany to debrief German rocket scientists, he understood the importance of their missile technology for carrying out the type of upper-atmosphere research that the NRL wanted to do. They created a Rocket-Sonde Research Branch in December of that year, and one month later the Army Ordnance Department invited the NRL and a consortium of Navy-funded scientists to begin a research program using sounding rockets.

Sounding Rockets

One of the most famous pen names in American literature is that of Samuel Clemens, who borrowed it ('I laid violent hands upon it without asking permission of the proprietor's remains') from a deceased Mississippi riverboat captain who wrote under that name in reporting river news for the New Orleans *Times-Picayune*. In order to ensure that the water was deep enough for safe passage, a crew member called the leadsman would toss a knotted rope weighted by a piece of lead over the side. When the water's depth exceeded two knots – 12 ft, or 3.65 m – he would call out 'By the mark, twain', or simply 'mark twain'.

The lead weight would sometimes be hollowed out at the bottom to collect samples of the riverbed material, to determine when it was changing from safe mud to hazardous rocks. This sampling process was, and still is, known as 'taking a sounding'. The French word *sondage* (from *sonder*, to probe) is used for the process of taking a sample – in modern usage it describes an opinion poll and further back in time it described a trench dug by archaeologists to inspect the stratigraphy of a dig site. Early in the twentieth century the word was applied to instruments, known as radiosonde devices, which are sent aloft on weather balloons to sample the atmospheric conditions and radio the data back to the ground. When small rockets started to be sent up on brief missions to measure conditions in the upper atmosphere, and eventually further up into the ionosphere and even further out to the altitudes that would later be used by orbiting satellites, they were called 'sounding rockets' to identify their purpose for making measurements while at altitude.

After the end of the Second World War hundreds of railcar-loads of captured v-2 missile parts were shipped to White Sands Missile Range (WSMR) in New Mexico and to the nearby Fort Bliss army base, along with a large group of German rocket scientists and engineers (illus. 64). Dozens of v-2s were launched from WSMR in the late 1940s and into the early 1950s, while a simultaneous programme to develop U.S.-made Aerobee rockets was also carried out. In both cases, the rockets were used largely for upper atmospheric and solar studies, with gradual improvements such as parachute recovery systems and solar pointing being added, as well as increased payload size for the series of Aerobee rockets.[18]

There are numerous small sounding-rocket programmes worldwide, in Europe, Asia and Australia, among others. The largest is in the U.S., where the main location of NASA's sounding-rocket launches is WSMR; some launches are also carried out at Wallops Island, Virginia, and at Poker Flat, Alaska. Sounding rockets offer some definite advantages over satellites: they are far less expensive to build and launch; the turn-around time from the start of a programme to launch is typically shorter; there are more opportunities for launch with sounding rockets (mainly due to the lower cost); and the programme offers a way to test new instruments and to train young scientists and engineers. On the negative side, the launch involves a solid-fuel rocket motor that induces fearsome acceleration (the rocket reaches supersonic speed in about four seconds!) and enormous vibrational loads. In addition, instead of the months and years of observing that are possible with an orbiting satellite, the amount of time in a sounding-rocket flight is typically limited to about five minutes. That amount, however, is infinitely larger than zero, so it remains a highly desirable way to do research. The availability of rocket observations is especially

64 A group of over 100 mainly German rocket scientists and engineers at Fort Bliss, Texas, in 1946, part of Operation Paperclip. Wernher von Braun is in the first row, seventh from the right. He and his team were moved to Huntsville, Alabama, in 1950 and there developed the Redstone and other rockets.

relevant to solar physics, because the solar corona and its dynamics are best studied from above our atmosphere at wavelengths far outside of the visible range. On 24 February 1949, a rocket launched from WSMR reached a record altitude of 250 km, becoming the first known man-made object to reach outer space.

The X-ray Sun

The existence of the solar corona was one of the greatest unsolved mysteries in all of astrophysics for about eighty years, from the first applications of spectroscopy to the corona at eclipse in 1860 until 1941, when the great Swedish physicist Hannes Alfvén reviewed the available evidence and concluded that the corona is extremely hot. The problem was that the corona seemed to be physically impossible. It was seen to have a spectrum very much like the photospheric surface of the Sun, with a temperature of 5,800 K. This is hot, to be sure, but not hot enough to allow the corona to extend out as far as is seen. The extent of an atmosphere is determined by a balance between its temperature, which tries to make it expand outwards, and gravity, which pulls it back down. An atmosphere at 5,800 K under the influence of the Sun's strong gravitational pull should extend only a tiny fraction of a solar radius out, not nearly as far as is observed to be the case. So explaining the 'impossible' extent of the observed corona was a major puzzle.

In addition, as we saw in Chapter Six, bright spectral lines were detected in the corona, emitting at very specific wavelengths, and none of those emission lines could be identified. They did not seem to correspond to any known elements! Many solutions were proposed, including a new element, coronium (in analogy with the newly discovered element helium, first identified in the solar spectrum). None of those solutions worked. Finally, a combination of laboratory studies on the spectroscopy of high-temperature plasmas, combined with detailed atomic physics calculations and with observations of certain peculiar variable stars, all came together to provide the answer. Those strange emission lines were found to be coming from exotic states of ordinary, known elements (such as iron and calcium) but with many electrons stripped away because of the exceedingly high temperatures. Collisions between the atoms of the hot coronal gas knock electrons out of their otherwise neutral atoms, and the higher the temperature, the more violent the collisions and the more electrons can be knocked loose by the collisions. Those free electrons eventually recombine with their partner ions, emitting a photon in the process. Meanwhile, other electrons from other atoms are knocked loose, in a continuing cycle of ionization and recombination.

These peculiar highly ionized (that is, with electrons removed) states of matter, combined with other more subtle spectroscopic evidence (such as the broadening of the observed spectral lines), together argued that the coronal gas is at an extremely high temperature, on the order of a million degrees K. The quasi-photospheric 5,800 K spectrum seen in the corona at solar eclipses turned out to be due to the extremely bright photospheric light scattering off of those electrons in the corona, providing a strong background of light at wavelengths around those of the emission lines, with the Fraunhofer absorption lines obliterated by Doppler shifts due to the rapid motions of the electrons, and making it seem misleadingly as if the corona is at the photospheric temperature.

The corona is so hot that its primary emission, the majority of the light that it emits, is at extremely short wavelengths, in the EUV and soft x-ray part of the electromagnetic spectrum. The first image of solar x-rays from the bright, hot, active regions in the corona was obtained in 1960 by the sounding rocket experiment of Herbert Friedman of NRL. The imager was a fairly crude pinhole camera, but techniques for focusing x-rays to form an image progressed rapidly in the ensuing decades, with sounding rocket instruments leading the way. An example of the high quality of the data that can now be

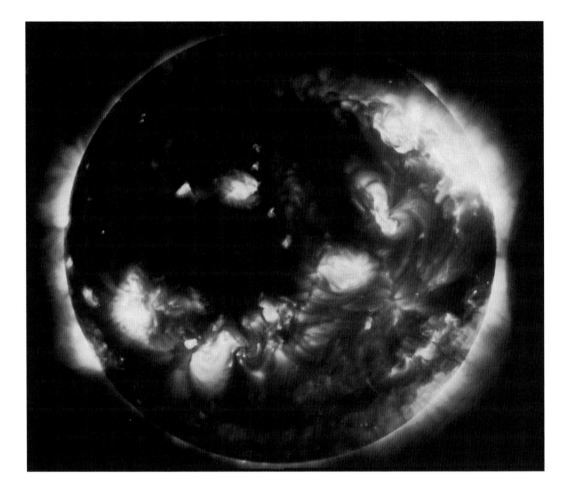

obtained with sounding rockets is shown in illustration 65. This image was recorded on 11 July 1991, at a time corresponding to a total eclipse of the Sun that was observed elsewhere. With fortuitous timing, full totality was observed directly over the Canada-France-Hawaii Telescope in Hawaii at exactly the moment when the Moon was just approaching the Sun over the rocket range at White Sands, New Mexico. The image was made possible by the use of a special coating deposited onto the mirrors of the Normal Incidence X-ray Telescope (NIXT) to make them reflective at soft x-ray wavelengths.

65 This photo shows a high-resolution image of the solar corona at soft x-ray wavelengths, taken from the NIXT sounding rocket on 11 July 1991. A crescent of obscuration from the Moon can be seen on the extreme right, and at exactly this moment a total eclipse was seen over Hawaii.

This allowed an image of the high-temperature solar corona to be obtained while the approaching dark disc of the Moon could be seen occulting part of the extended corona.

An even more impressive rocket observation, using the same type of reflective coating, was obtained exactly 21 years later, on 11 July 2012, by the High Resolution Coronal Imager (Hi-C). Illustration 66 shows a comparison between an EUV image of the corona taken at 19.5 nm with almost the same spatial resolution as the NIXT image, using NASA's Solar Dynamics Observatory, and the simultaneous Hi-C rocket image, with the resolution about five times higher. Such data, as well as data obtained from new types of spectrometers, and from imagers at other wavelengths from ultraviolet to hard x-rays, have proven to be a major component of solar research. Because it provides a relatively quick and low-cost means of putting an instrument into space, the sounding rocket programme is a vital part of our efforts to understand the Sun and,

66 Frame from the High Resolution Coronal Imager (Hi-C) sounding rocket flight, comparing the detail observed in Hi-C, with its improved resolution, to that seen in NASA's Solar Dynamics Observatory (SDO). The two images were obtained at the same wavelength and at the same time.

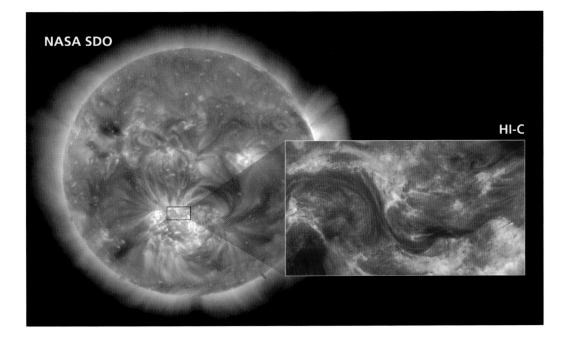

by extension, all areas of astrophysics where hot magnetized plasmas are observed.

Satellites

History was made on 4 October 1957, when the Soviet Union launched Sputnik 1, the world's first artificial satellite, thereby marking the start of the Space Age – and setting off the u.s.–ussr space race. Led by Sergei Korolev, whose name was kept secret at the time, the Soviet team launched an 83-kg polished metal ball that was visible by eye from the ground near sunrise and sunset as it passed overhead, and that emitted a continual beeping radio signal that was detectable by amateur radio receivers. The u.s. had been working for several years to put a satellite into orbit, under the u.s. Navy Vanguard programme. Vanguard was attempting to launch a 1.6-kg payload but the launches, broadcast live on tv, kept exploding spectacularly just seconds after ignition. Even more dramatic was the launch of Sputnik ii one month later, with a heavier payload and with a dog named Laika as a passenger.

The launch of Sputnik caused near panic in the u.s., leading to a furious effort to launch a satellite, the program now under the leadership of Wernher von Braun and his Jupiter-C rocket. On 31 January 1958 Explorer 1 was launched, and thereafter the programme continued with a series of lightweight, scientifically useful satellites. Among other achievements, these experiments – and a similar series in the ussr – discovered and explored the Van Allen radiation belts around the Earth, which are extended torus-shaped regions of high-energy charged particles trapped by the Earth's magnetic field. On 1 October 1958 the National Aeronautics and Space Administration (nasa) was created, and funding for education, especially in maths and science, was sharply increased.

All of this activity provided a great benefit to scientific research, especially via the civilian agency nasa. A vigorous programme of small satellite launches began, not only orbiting the Earth, but visiting the Moon and Venus in the 1960s. Studies of high-energy solar particles known as cosmic rays and lower energy particles known as solar wind benefited from the programmes, and a series of eight small satellites named the Orbiting Solar Observatory (OSO) was launched between 1962 and 1971 to study the Sun at ultraviolet and x-ray wavelengths.[19] After the Apollo programme of Moon landings was terminated, a remaining Saturn V rocket was used to launch Skylab, the first U.S. space station, followed by three separate periods of astronaut visits launched on Saturn IBS. In place of the *Apollo* lunar module a set of solar telescopes, named the Apollo Telescope Mount (ATM), was installed and operated by the Skylab astronauts, who retrieved and returned large film canisters to the scientists waiting on the ground. (The ATM is now on public view at the Smithsonian Institution's National Air and Space Museum in Washington, DC.)

67 The terrestrial atmosphere is opaque to most electromagnetic wavelengths outside of the visible and a portion of the radio spectrum. Some narrow windows exist for transmission of visible, infrared and radio wavelengths through our atmosphere to the ground.

The wavelengths of the electromagnetic spectrum that we call visible light comprise only a very tiny slice of the spectrum, as shown in illustration 67. The coloured vertical rainbow in this drawing shows where the visible colours fall, while shorter ultraviolet wavelengths are to the left and longer infrared and radio wavelengths are to the right. The wavy black line in the figure indicates the opacity of Earth's atmosphere to each wavelength, where 'opacity' means how much of that wavelength gets absorbed before it reaches us on the ground. Thus 100 per cent means that the atmosphere is totally opaque at that wavelength, so all of it is absorbed and none of it reaches us. Visible light is therefore one of the few wavelength bands that gets through.[20] Ultraviolet wavelengths are heavily absorbed high in the atmosphere, which is fortunate for us because they are harmful to life, but unfortunate for observational astronomy because it means that we need to get our detection apparatus above the absorbing atmosphere. Infrared wavelengths are partially transmitted in a few narrow wavelength bands and can be seen from mountaintop observatories, high-altitude balloons or specially equipped aircraft, while a broad swathe of the radio spectrum can be detected on the ground. Other than these few special circumstances, detection of all the other wavelengths requires that we put our detectors above the atmosphere into space.[21]

As we move through the spectrum from visible light to ultraviolet (UV) to the Extreme UV (EUV) and to x-ray wavelengths, by taking images through filters at shorter and shorter wavelengths, we detect progressively hotter parts of the solar atmosphere (illus. 68). At first we see visible-light sunspots, or with the methods described in Chapter One we can measure the magnetic fields on the solar surface at visible wavelengths. Using blue wavelengths we see slightly higher up than the photosphere, to a region called the 'temperature minimum' where, as the name implies, the temperature is lower than that of most of the photosphere and also lower than that of most of the chromosphere. At the higher

68 The atmosphere of the Sun is highly structured not only spatially but in temperature: wavelengths that show different temperatures will show different parts of the solar atmosphere. For this reason, NASA's SDO contains multiple telescopes operating at different wavelengths, so that together they can form a complete picture of the atmosphere.

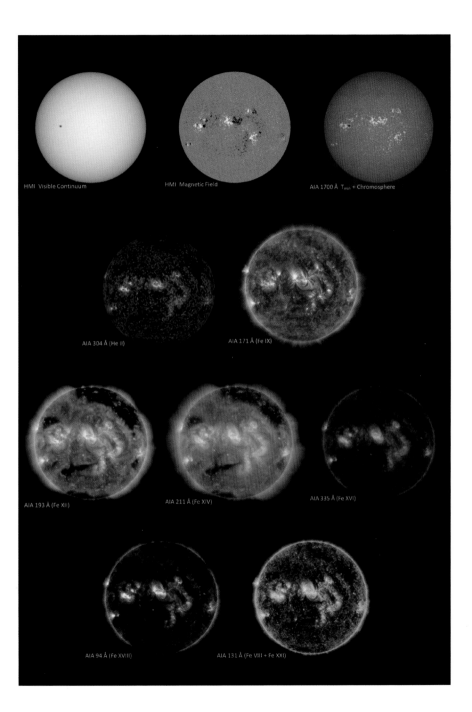

temperatures it makes very little sense to speak about 'layers' of the atmosphere because, as we saw in earlier chapters, the chromosphere is highly structured, being made up of spiky vertical spicules, segments of loops and other three-dimensional components. The same applies as we go to even shorter wavelengths, where loop-like plasma structures dominate, constrained and defined by the magnetic fields that permeate the atmosphere. Throughout, we continue the method started by Hale of selecting narrow wavelength regions around a particular spectral line, so that we obtain an image of the regions that are producing that wavelength. In the EUV, those regions are at temperatures of several million kelvins.

Despite more than seventy years of effort, there is no widely accepted theory of how the corona is heated to such enormously high temperatures. But one major fact is clear: the corona is brightest and hottest where the strongest magnetic fields emerge from inside the Sun. When viewed in visible light, we see sunspots at the locations where strong fields pop through the surface (see Chapter One). But if we block the visible light and instead image the EUV and x-ray emission, we see that the atmosphere about the sunspots is filled with complex structures that are clearly connected to the magnetic field and which are so hot that the light they emit has shifted down into those short wavelengths. With modern ground-based instruments, and telescopes in space, we are now able to see this connection clearly.

69 The close connection between the strong magnetic fields at the solar surface and the hot, x-ray-emitting plasma above the surface is shown in this image from 8 August 2015. The three panels show, respectively, the surface magnetic field in the vicinity of a sunspot region, the calculated magnetic field above the surface, and the observed coronal structure.

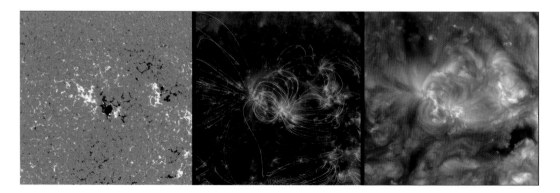

Illustration 69 shows a comparison between the magnetic fields measured on the solar surface and the coronal structures seen above the surface in EUV images. The leftmost panel shows a portion of a magnetogram, with black-and-white patches indicating where the magnetic field emerges out of the surface and where it re-enters the surface. At the centre of this image there is a strong black patch with a circular part embedded: this is a place where a large sunspot is located. To the left and slightly above we find the corresponding opposite polarity (white) magnetic field region. To the right of the spot there is another concentration of white polarity, a complexity that made this active region highly productive in solar flares. Further to the right we see another bipolar, black-and-white region. This one is older and the fields are less concentrated; the region is also less active.

We do not yet have a reliable method for measuring the magnetic fields in the corona, but we can estimate the field configuration above the surface by extrapolating mathematically from the measurement of the field at the surface. There are a number of different ways in which such an extrapolation can be done, and the simplest of them is shown in the centre panel of illustration 69. We see that there is a spray of fields diverging out of the sunspot, implying that its strong magnetic fields spread out in all directions from the dark umbra. We further find that the sunspot is connected by loop-shaped magnetic fields to the white polarity to the left of the spot. This is consistent with our image of a horizontal rope of magnetic field emerging from inside the Sun, with fields coming out at one end of the omega-shaped bundle and returning to the surface at the other end. We see also that this region has developed connections to the older region to the right, with magnetic loops connecting the two.

Examining the EUV image in the right-hand panel of the figure, we see that the hot coronal plasma, which shows the outlines of the magnetic field because it is constrained to follow the field direction,

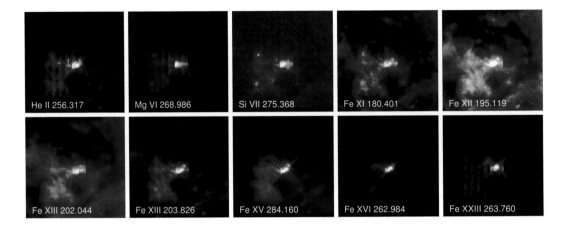

| He II 256.317 | Mg VI 268.986 | Si VII 275.368 | Fe XI 180.401 | Fe XII 195.119 |

| Fe XIII 202.044 | Fe XIII 203.826 | Fe XV 284.160 | Fe XVI 262.984 | Fe XXIII 263.760 |

exhibits the same connectivity between opposite polarity regions as is predicted by the magnetic field extrapolation.

In terms of their effect on the Earth and other solar system objects, one of the circumstances that we are especially interested in understanding is the sudden release of energy that has been stored in the corona during the evolution of a magnetic region. One consequence of this energy release is a solar flare, as shown by the small but intense bright spot in the active region imaged in illustration 70. (A related phenomenon, the coronal mass ejection, will be discussed in the next chapter.) In a flare, the corona becomes rapidly brighter and hotter in a localized region, reaching temperatures of tens of millions of degrees and briefly outshining the entire Sun in x-ray light.

Observation from space allows us to see that the corona in and around active regions contains material at flare temperatures even in the absence of any obvious flares. The NASA NUSTAR satellite, designed to focus and detect high-energy x-rays for studies of exotic objects throughout the universe such as black holes and relativistic jets, has also been used to look at these high-energy x-rays from the Sun. Illustration 71 shows such an image from NASA/Caltech's NUSTAR satellite, overlaid on a lower-energy (about 1 kilovolt)

70 An X-class flare. This sequence of images from the EUV Imaging Spectrometer (EIS) on *Hinode* shows progressively higher-temperature parts of the Sun's atmosphere. The upper left corner shows material in a relatively low-temperature part of the atmosphere at about 70,000 K – and the images progress ever upwards in temperature through the heart of the flare up to over 15 million K in the image at the lower right. Each image shows a narrow swatch of the flare, which together can be combined to create a three-dimensional picture.

71 X-rays light up the Sun in this image containing data from NASA's Nuclear Spectroscopic Telescope Array, or NUSTAR. The high-energy x-rays seen by NUSTAR are shown in blue, while green represents lower-energy x-rays from the X-ray Telescope instrument on the *Hinode* spacecraft. NUSTAR data shows x-rays with energies between 2 and 6 kiloelectron volts; the *Hinode* XRT data has energies of 0.2 to 2.4 kiloelectron volts; and the Solar Dynamics Observatory data, taken using the Atmospheric Imaging Assembly instrument, shows extreme ultraviolet light with wavelengths of 171 and 193 Å.

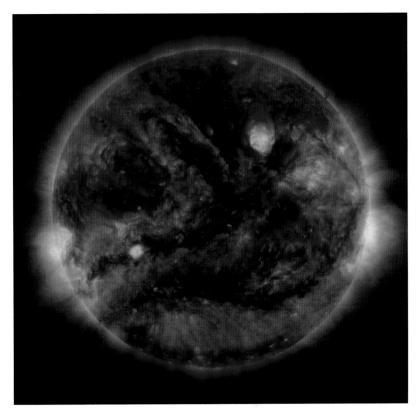

image from the *Hinode* spacecraft's XRT imager. The bluish glow in and above the XRT active regions shows the presence of 2–6 kilovolt x-rays, indicating the presence of a process that is producing flare-like high temperature plasma. It is at present a rather mysterious and unexplained phenomenon.

In order to study the way in which energy is stored and then released in the corona, we need to be able to observe the long-term evolution of the magnetic fields and the hot plasma intertwined with them. With satellites in space observing the Sun nearly continuously, we are now able to follow the evolution of the corona over a period of months and even longer. Illustration 72 shows nine solar rotations, arranged left to right and top to bottom with images taken

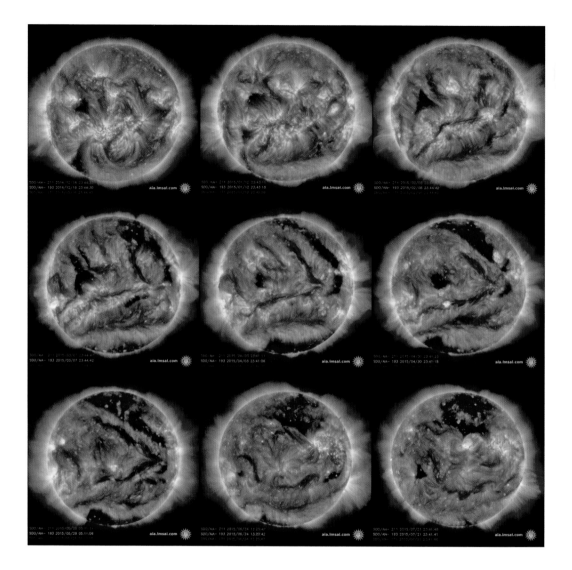

72 The development of the large-scale corona over a period of nine solar rotations is seen in this series of EUV images. They show the same part of the Sun's surface, month after month, reading left to right and top to bottom. The way that the differential rotation of the Sun affects the corona, the atmosphere above the visible surface, is seen in this series of images taken over a period of eight months. The more rapid rotation of the solar equator relative to its poles stretches the structures out into V shapes, because the mid-latitude features are being pulled around more rapidly, leaving the high-latitude features lagging behind.

27 days apart so that the same side of the Sun is facing us in each image. Because the Sun rotates with the equator moving around more quickly than do the poles, as discussed in Chapter Two (illus. 15), features near the equator recur in about the same place after 27 days but those at higher latitude gradually get left behind. This differential rotation leads to V-shaped patterns with the equator at the point of the 'V'. We can also see the spreading out of coronal structures across the solar surface, as regions that were bright and compact when they first emerged become fainter and more diffuse as their footpoints – the roots of their magnetic fields anchored in the solar surface – are pushed around by the surface convection and become spread across the surface. Both of these behaviours, differential rotation and turbulent diffusion, were central elements of the Babcock–Leighton dynamo model that we discussed in Chapter Three, which has now been shown to be both prescient and largely accurate.

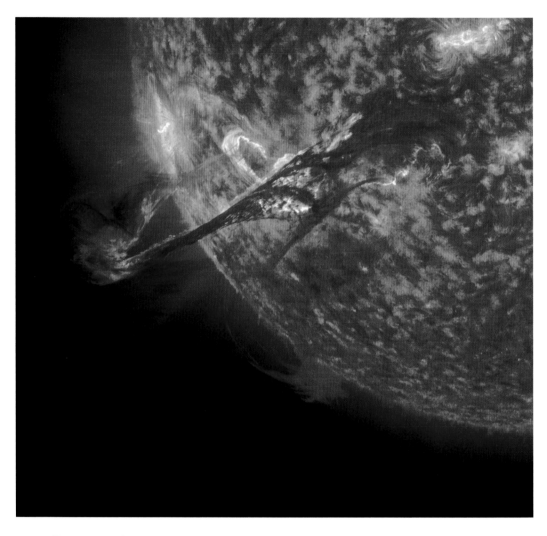

73 Large filament erupting from the solar surface out into interplanetary space.

EIGHT

Storms from the Sun: A Discussion Mostly About Particles and Fields

Achieving an understanding of the connection between events on the Sun and disturbances reaching the Earth took more than two centuries. It involved scientists, philosophers, mathematicians, the militaries of powerful nations and eccentric inventors, to name a few, and required tying together the behaviour of compass needles, discussions about statistical methods, observations of aurorae and sunspots, proposals about mysterious M-regions, and speculations about comet tails. Disagreements, arguments and counter-arguments raged among some of the greatest scientists in the world, with inspired guesses and vicious criticisms, and many advances and setbacks. The question was finally settled only by direct observation from interplanetary space, which dramatically demonstrated how the seemingly impossible connections did in fact exist. A list summarizing some of the major steps along the way to our present understanding is provided in Table 1.

As early as 1515, Thomas More had warned of the dangers of relying on new technologies, saying of island natives who had acquired a magnetic compass from European sailors:

> They sailed before with great caution, and only in summer time, but now they count all seasons alike, trusting wholly to the loadstone, in which they are, perhaps, more secure than safe.

The magnetic needle was indeed tricky to use. We discussed earlier the discrepancy between magnetic north and geometric north: the Earth's magnetic North Pole is not located at the same place as the North Pole that is determined by the Earth's axis of rotation. So it is necessary to produce maps showing the difference ('declination') between the two across the surface of the Earth. Producing such maps would suffice for navigation, as do charts of water depth, reefs and land masses. But the location of the magnetic North Pole is drifting noticeably on fairly short time scales, and the rate of variation of the declination varies considerably from one place to another on the Earth's surface, so that maps of the declination need to be updated often, every few years if accurate navigation is required. Worse, it was also found that the compass needle shows a variation every day, drifting slightly as the Sun comes up and drifting back in the evening. Even worse than that, there were occasional days when the needle would flicker erratically and become useless for navigation, as it would swing wildly back and forth.

The rapid drift of the magnetic North Pole (illus. 18) and the complex movements of the magnetic compass needle were only the beginning of the difficulties. Complex models of the Earth's magnetic field had to be introduced when the simplest ones failed, as happened with Gilbert's simple dipole model, precipitating Halley's more complicated model. Tantalizing connections between seemingly disparate phenomena, such as aurorae and sunspots, led to arguments over statistical analysis methods and the development of new ways to evaluate causal connections statistically. Great debates raged over whether and how the Sun could produce the proposed effects at Earth, given that no visible connections could be found between the two. New instruments were built and new theories were developed. In effect, every aspect of what we think of as the 'scientific method' entered the fray.

Table 1

1724   Clockmaker George Graham invents a sensitive magnetic needle that can detect slight changes in the Earth's magnetic field; finds diurnal variations.

1741   Anders Celsius and his student O. P. Hiorter show that the magnetic variations are accompanied by aurorae; works with Graham to show that the effect is widespread, not local.

1843   Heinrich Schwabe publishes sunspot records, showing that spot numbers increase and decrease regularly in ~ten year cycles.

1851   Alexander von Humboldt publicizes Schwabe's data in his treatise *Kosmos*.

1852   Edward Sabine correlates the sunspot cycle with the rate and size of magnetic disturbances.

1859   Richard Carrington sees a white-light flare, notices ensuing prompt and delayed magnetic disturbances and aurorae.

1878   Balfour Stewart proposes that diurnal variations of the compass needle are caused by upper atmospheric (ionospheric) currents.

1892   Hale invents the spectroheliograph, photographs the brightening on the sun from a solar flare.
William Ellis demonstrates a statistical link between geomagnetic variability and solar cycle.

1898   Ellis shows a strong correlation among sunspot numbers, strength of the daily magnetic variations and frequency of magnetic storms over five full solar cycles.

1904   Walter Maunder shows a 27-day recurrence period for geomagnetic storms; shows Annie Maunder's 1898 eclipse photo with coronal 'rays'.

1905   Another RAS presentation by Walter Maunder, at which Joseph Larmor proposes electron beams to carry disturbances from the Sun to Earth.

1908   Kristian Birkeland proposes solar currents (now called auroral electrojets), induced by 'cathode rays'; ridiculed by Sydney Chapman.
Hale shows that sunspots are magnetic regions.

1919   Chapman argues that diurnal magnetic variations are due to solar UV radiation.

1929   W.M.H. Greaves and H. W. Newton show that large geomagnetic storms are associated with sunspots, while small storms show a 27-day recurrence period, without spot associations.

1932   Julian Bartels analyses 27-day recurrent storms; shows lack of sunspot correlation; denotes solar source as 'M regions'.

1933   Chapman and Ferraro propose 'magnetic cloud' model of magnetic storms, including 'Chapman–Ferarro cavity' around Earth, i.e., the magnetosphere.

1951   Ludwig Biermann analyses comet tails; proposes stream of particles from the Sun to explain why the tails always point away from the Sun.

1958   Gene Parker proposes that the high temperature of the corona will lead to its continual expansion in a supersonic solar wind; ridiculed by Chapman.

1962   Interplanetary spacecraft *Lunik 1–3* and *Mariner II* detect a supersonic solar wind.

1973   Coronal holes identified as the source of recurrent high-speed streams.
Coronal transients, soon to be renamed coronal mass ejections, identified as magnetic clouds and sources of geomagnetic storms.

Magnetic Storms

Throughout the first half of the nineteenth century the most famous, celebrated and sought-after scientist in the world was Alexander von Humboldt. With enormous energy he explored a wide variety of subjects, effectively inventing the concept of an ecosystem and exploring the complexities of the natural world in ways that inspired Darwin later in the century. Among von Humboldt's many interests was the puzzle of terrestrial magnetism and its variations, so he included sensitive magnetic instrumentation in his voyage to South America of 1799, obtaining thousands of readings up to the time of his return to Berlin in 1804, where he continued the magnetic observations. After noticing the connection between aurorae and disturbances of the magnetic needle, von Humboldt coined the name 'magnetic storm' for the phenomenon, and we continue to use that designation to this day. He moved to Paris in 1807 and spent the next twenty years writing up his results.

Von Humboldt attacked the magnetic storm problem with his typical energy and depth, enlisting the help of eminent scientists such as Carl Friedrich Gauss and Wilhelm Weber to join him in trying to understand the nature and source of the Earth's magnetic field and its disturbances and convincing the Prussian court to fund a network of magnetic observing stations throughout Europe and across Russia. In order to extend the network further, he worked surreptitiously with his British counterpart, Edward Sabine, to put magnetic stations throughout the British Empire.

Observations continued for decades, with data pouring in from across the world and Sabine charting variations in an attempt to find a pattern. Finally, in 1850 von Humboldt published his massive work Kosmos, which included a description of Schwabe's discovery of the sunspot cycle. Sabine examined the sunspot data (after his wife, Elizabeth Leeves,who was translating Kosmos, called it to his attention) and found two major correlations: the sunspot numbers tracked his charts of magnetic storms perfectly, and the strength of the daily variations of the compass needle tracked the average sunspot numbers during each cycle. And so began a century-long argument over the possibility of, and nature of, the connection between solar activity and terrestrial disturbances.

Although not evident at the time, a major piece of the puzzle was found in 1859 when Richard Carrington happened to be recording sunspots exactly at the time that a rare white-light flare occurred. Only the very largest and most intense solar flares can be detected in white light without using special narrowband wavelength filters, so this event was exceptional and the accompanying terrestrial disturbances will be discussed later in this chapter. One of the observations that Carrington examined was the magnetic record at the nearby Kew Observatory, part of the magnetic observing network established by Sabine. Kew reported a short-lived disturbance at the time of the flare, and then a huge, long-lived magnetic storm eighteen hours later, indicating that two different phenomena associated with the flare had occurred, one propagating at roughly the speed of light (at which speed light from the Sun reaches Earth in just over eight minutes) and the other at the much slower pace of 'only' 5 million miles per hour.

But the question of prompt vs delayed effects only led to confusion, and uncertainty over whether or not this correlation was merely a coincidence actually hindered progress in understanding the meaning of the observations. Still, the statistical correlations between sunspot numbers and magnetic storms persisted. In 1880 William Ellis published an extended study confirming the earlier work of Sabine and others, showing a correlation between the size of the daily compass-needle variations and the average sunspot number. In 1892 he published another study showing that the onset of geomagnetic storms worldwide was effectively simultaneous, starting at the same time at all locations within a fraction of a minute, and he suggested that something external to the Earth was causing the storms.

The search for a mechanism that could explain how the Sun could affect the Earth in this way was dealt a major setback in 1892 by the president of the Royal Society, Lord Kelvin, who presented to the Society a calculation of how much energy would be needed for the Sun 'as a magnet' to produce the effects seen at Earth. Using a model in which the entire magnet of the Sun suddenly changed strength, he calculated that it would require a full four months' worth of the total solar energy output to produce the magnetic effects lasting only eight hours at Earth, leading him to conclude that the observed correlations were all 'a mere coincidence'. The problem with his calculation, as it turned out, is that he was using an incorrect model of the phenomenon.

The year 1898 was a fairly ordinary one by historical standards, including among its notable events the merging of Brooklyn to New York City to create the five boroughs known today, the annexation of Hawaii to the u.s. and the acquisition by the u.s. of Puerto Rico, Guam and the Philippines following the Spanish–American War. Within the more limited realm of solar-terrestrial physics, however, there were significant advances made by William Ellis and a married couple, E. Walter and Annie Russell Maunder.

Edward Walter Maunder came from an underprivileged background, attending King's College, London, and working in a bank to pay for his education. Although he never completed his studies and did not earn a degree, a civil service reform allowed him in 1873 to pass an exam for a job as an assistant at the Greenwich Observatory. The Astronomer Royal, George Biddell Airy, had recently won a political struggle to take on the regular solar observations that were no longer being done at the Kew Observatory, and needed increased staff for the extra work. Maunder's job was to take routine daily photos and spectra of the Sun and its spots. Maunder's working situation improved when the curmudgeonly Airy retired in 1881 and his assistant, William Christie, became the new Astronomer Royal. Christie quickly upgraded the importance of solar observations within the Observatory's tasks and made Maunder the head of the solar department. In 1891 he was granted an assistant and, following his long-held belief that women should have professional standing in science, hired a young Irish mathematician, Annie Scott Dill

Russell, a graduate of Girton College, Cambridge. It was the first time that any woman had been employed by the Royal Observatory.

Annie may have started as a 'computer' but she turned out to be a capable astronomer, taking on a large share of the data-gathering and recording, and inventing a wide-field camera funded in part by her old Cambridge college. The two married in 1895 – which meant that she had to quit her job, since married women were not allowed to work at the Observatory, or throughout the Civil Service for that matter. They continued to work together, and decided that Annie's camera could be used to photograph a solar eclipse, due over India on 22 January 1898. Although fairly late in the sunspot cycle, so that activity levels were reduced from the maximum, the corona turned out to be exceptionally bright at the eclipse, and the camera caught spectacular views showing greatly extended radial structures emanating from the solar disc. The longest of the streamers was detected as far as 6 million miles from the limb of the Sun, inspiring Walter to believe that they might have found an answer to Kelvin's objection. Instead of radiating out its magnetic influence uniformly in all directions, as Kelvin had assumed, perhaps the magnetic influence was somehow channelled into narrow beams of energy. This would have a huge effect on Kelvin's calculation, greatly reducing the total amount of energy needed. It would also explain why some solar events caused disturbances at Earth and others didn't, depending on whether the beam was or was not directed towards Earth.

As it happens, the streamers that the Maunders recorded at the 1898 eclipse are not the sources of geomagnetic storms, but they did put Walter on the track towards a more convincing argument, one that required eight years to work out. Meanwhile, William Ellis followed up on work he had reported earlier on the statistical connection between sunspots and geomagnetic storms, work that had been pushed aside by Lord Kelvin's strong objections. Ellis extended his previous work to include the entire period from 1841 through

1896, covering five solar cycles. The result was evident: the average sunspot number over the course of the cycle and the strength of the daily geomagnetic variations tracked each other perfectly. Though the correlation was clear, Ellis followed a theory popular at the time and concluded that some unknown third cause was influencing both the Sun and the Earth.

A key moment occurred for the Maunders when they realized that after a large magnetic storm there was a marked tendency for another storm to occur 27 days later, a time delay that could only be associated with the rotation period of the Sun's active region longitudes. It was 1904 before Walter Maunder was ready to present his analysis to the Royal Astronomical Society (RAS), relying heavily on a statistical association from a 'periodgram [sic]' analysis, an early form of the modern technique for discovering the recurrence frequencies in time series of data. His presentation, which also included Annie Maunder's 1898 photo and a mention of the Swedish scientist Svante Arrhenius's suggestion that charged particles might be emanating from the Sun, led to a protracted discussion and to an agreement to return to the topic a few months later, in 1905.

The 1905 RAS discussion was vigorous, but led to a grudging acceptance of the reality of a 27-day occurrence period, corresponding to events originating within narrow, well-defined longitudes on the Sun that remain in place for several months at a time. The expert on periodgrams, Arthur Schuster, ended up conceding that Maunder's analysis might be correct, and a visitor, the eminent physicist and Lucasian Professor Richard Larmor, spoke up to agree that the statistical analysis was indeed correct and also to cite recent work showing that a stream of electrons could carry electromagnetic energy from one place to another.

This so-called corpuscular hypothesis was vigorously pursued by the Norwegian scientist Kristian Birkeland, who established a network of auroral observatories in Norway and found a global pattern of electric currents in the polar regions that were producing the magnetic disturbances. He carried out laboratory experiments using a magnetized sphere bombarded by electrons in a chamber, showing that rings of light were produced near its poles, just as on Earth, and proposed that charged particles crashing down into the polar region were responsible for the auroras. His 1908 proposal that the deflection of the magnetic compass during auroral storms was due to electric currents flowing along magnetic fields connecting what is now called the magnetosphere to a layer of the upper atmosphere now called the ionosphere was greeted with scorn and ridicule. It was not until 1967 that satellite measurements confirmed the existence of what are now called Birkeland currents. Hannes Alfvén, whose own work was treated harshly despite his Nobel Prize, was one of the few to champion Birkeland's ideas.[22]

Gradually, it became clear that there are two major classes of magnetic disturbance: recurrent and sporadic. The sporadic events were found to be generally associated with sunspots, while the recurrent storms were classed as being due to unidentified 'M-regions' on the Sun, which were found to be associated with low-brightness regions in the corona. Now that we have the ability to observe the Sun from space and also to make direct local measurements of the solar wind in space,[23] the puzzles of the relationship between events on the Sun and disturbances at Earth have been solved. The two types of disturbances – sporadic and recurrent – come respectively from coronal mass ejections and coronal holes. Coronal holes are sources of high-speed solar wind, an outflow of the coronal matter from regions that have become open to interplanetary space so that the corona can escape, leaving those regions with a lower density of coronal matter and consequently low brightness. Coronal mass ejections are more closely associated with sunspot regions, but typically after those regions have fully emerged to the solar surface and have begun to disperse and spread out across the surface. Dense concentrations of coronal particles called filaments can collect in the magnetic fields of these regions, and may sometimes be ejected out into space, along with the magnetic field that supports the dense material (illus. 73).

Coronal Holes and the Solar Wind

Observations from the ground hinted that recurrent geomagnetic storms originate from low-brightness regions in the corona, and debates raged over the features to be identified as the source. The explanation for the puzzling connection between sunspots and recurrent geomagnetic disturbances, while at the same time noting that there is no direct day-to-day correlation between the two, turns out to be fairly simple: coronal holes are produced by the remnant magnetic fields from the decay of large sunspot regions. So they exist where large sunspot regions *used to be*, or more accurately where the magnetic field that used to be in sunspots has now got to.

Once observations from the ground and from space were combined, they showed the decay of strong magnetic field regions by the spreading out and dying away of sunspot magnetic fields and the related formation of coronal holes. It soon became clear that these large open-field regions of the solar atmosphere are produced when an extensive area of the solar surface becomes dominated by a single magnetic polarity, for instance from the trailing polarity of an active region spreading out across the surface. When a large portion of the surface has one magnetic polarity dominating, the magnetic field above that part of the surface all tends to line up pointing in the same direction, vertically oriented. Still, the magnetic field would remain close to the Sun and the field emanating from this part of the surface would turn back down and end up at another part of the surface if there were no corona present. But the coronal plasma, which had been kept close to the surface by the closed magnetic fields, pushes outwards from the surface because of its high temperature, adding to the tendency of the field to open

up and to point out into interplanetary space. Eventually, the
combination of the distribution of magnetic charges across the
surface tending to make the field vertical and the upwards pressure
from the coronal plasma acting together cause the field to open
up. These open regions are places where the coronal plasma is no
longer confined by the magnetic field but is instead free to expand
outwards and leave the Sun, in the form of high-speed streams of
solar wind. The outflowing wind brings the magnetic field along
with it, so the two together form long streamer-shaped structures
that extend out to great distances from the Sun. At the very end of
this book we will discuss how far out they go and where it all ends.

An example of the connection between surface magnetic fields
and the existence of coronal holes is shown in illustration 74. On
this date, 28 February 2015, there was a large coronal hole extending
up from the south polar region of the Sun towards the equator, a
roughly trapezoidal shape as outlined in the top half of the figure,
a three-colour coronal image from the Solar Dynamics Observatory
(SDO) Atmospheric Imaging Assembly (AIA). A magnetic field map
taken at the same time by the HMI instrument of the SDO shows
that this part of the solar surface is, on the large scale shown
here, dominated by one magnetic polarity, as indicated by the
predominantly black magnetic field polarity there, sprinkled
amid the grey non-magnetic background. Other places on the
Sun, such as the bright active region above and slightly to the left
(east) of the equator, show a bipolar magnetic field on the surface,
having both black and white polarities visible.

The difference between having a bipolar magnetic field on the
surface vs having a large unipolar region can be seen by using a
method that extrapolates from the observed surface magnetic field
to calculate what the field might look like in the space above the
surface. Above regions of the surface that have a strong bipolar
magnetic field, we find that the corona consists of structures that
are closed, going from one location on the surface, arching across

74 An SDO image shows
a large coronal hole
extending up from the
southern polar region (top),
while an SDO HMI magnetic
field map shows the
dominance of one magnetic
polarity (black in this
presentation of the data)
in the coronal hole region
(bottom).

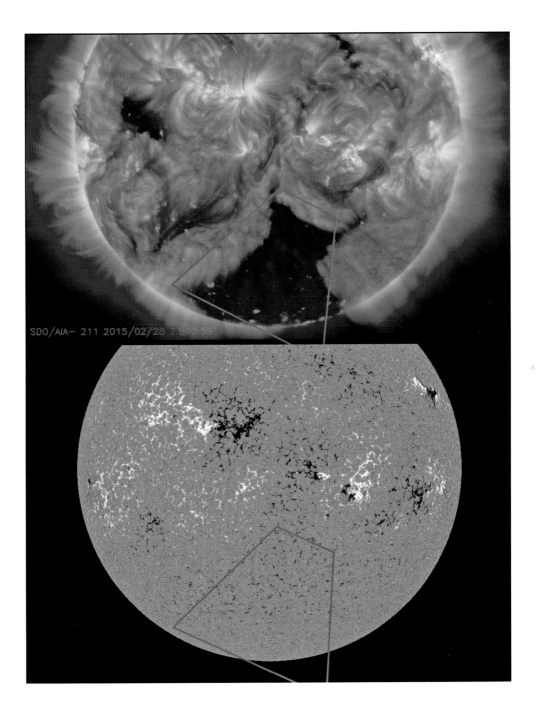

SDO/AIA- 211 2015/02/28 23:40:59

and re-entering the surface at another nearby location. One such
region was identified in the above discussion of illustration 74,
and the strong coronal emission from such closed field regions is
clearly seen in illustration 75, especially in a band along the equator
of the Sun where a chain of active regions stretches across the solar
disc from left to right. In contrast, the upper portion of the Sun on

the day that this image was obtained has an area similar to the one shown in illustration 74 where the magnetic field is more unipolar as the result of previous magnetic field emergence several months in the past. The result is that the magnetic field lines that we calculate tend to project outwards without returning to the solar surface: they can be seen projecting up, out of the frame of the image at the top, effectively opening out into interplanetary space (whether or not they remain technically closed is beyond our present ability to measure). This is what we mean by saying that coronal holes are open-field regions. Rather than confining the coronal plasma and holding it in place near the surface, as happens in closed-field regions, the coronal plasma in a coronal hole is free to expand outwards and leave the Sun, heading out into interplanetary space and possibly towards the Earth.

All of the planets in the solar system lie in nearly the same plane, roughly aligned with the equator of the Sun, and are moving around the Sun in the same direction as the solar rotation. So here at the Earth, we feel the solar wind that is in this equatorial plane, and until recently we could only infer, indirectly, the properties of the wind radiating out in other directions. This limitation was addressed directly by the *Ulysses* Mission, a joint ESA/NASA programme that launched a set of experiments to study the space environment out of the ecliptic plane, aiming to pass directly over the poles of the Sun. Moving a spacecraft out of the plane of the Earth's orbit requires a strong force and potentially an enormous amount of energy, but orbital change was accomplished with the help of the largest planet in the solar system, Jupiter. *Ulysses* was launched from the Kennedy Space Center on the Space Shuttle *Discovery* on 6 October 1990. Booster rockets then sent it *away* from the Sun and towards Jupiter, whose strong gravitational pull was used to fling *Ulysses* out of the ecliptic plane via a fly-by on 8 February 1992. The mission was successful beyond expectation, swinging completely around the Sun over each of its poles three

75 A large coronal hole is seen in this image of the million-degree corona on 19 June 2013. Overlaid on the image are lines showing the calculated direction and shape of the magnetic field that controls the hot coronal plasma. In the active regions near the equator the field is generally closed, beginning and ending at the solar surface, but in the coronal hole the field is open, beginning at the surface and extending out into interplanetary space, far past the orbit of the Earth.

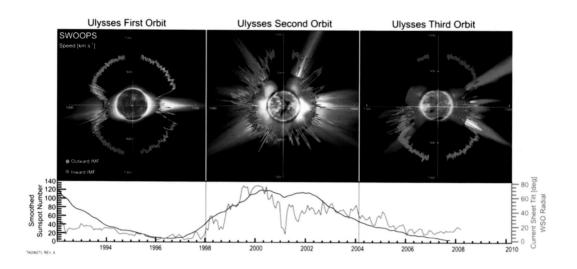

76 Solar wind speed measured from pole-to-pole around the Sun by the SWOOPS experiment on the *Ulysses* spacecraft over the course of three successive passes over both poles.

times, in an elliptical orbit with a six-year period, passing first over the south pole in 1994, then the north pole in 1995, then south to north again in 2000/2001 and a third time in 2007/2008. Finally running out of power from its onboard generator, the mission was declared over in 2009.

These three passes corresponded to a solar minimum, then a solar maximum and then another minimum, and the character of the solar wind changed dramatically from one pass to the next. The images in illustration 76 show a jagged line surrounding the Sun and a crossed pair of axes that indicate the measured speed of the solar wind at each latitude, determined as the spacecraft moved around the Sun from south pole to equator to north pole and down again. The most obvious effect seen in travelling around from pole to pole is that the solar wind speed is low near the equator (the jagged line is close to the Sun) and higher near the poles (the jagged line is further away from the Sun). So the closed-field regions near the equator seem to allow a solar wind, but at fairly low speeds, while the coronal hole regions at the poles are sources of high-speed solar wind. This much is seen at solar minimum, in the left and right panels. The middle panel, taken at solar maximum, shows a different pattern. The Sun at this phase of the sunspot cycle is covered by so many active regions that closed regions and the streamers above them are dominant, with almost no sign of polar coronal holes except for a small region at the very north. The solar wind in turn is dynamic and variable, generally of moderate speed, with a bit of high-speed wind at the highest northern latitudes.

Coronal Mass Ejections

Coronal holes and other open-field regions in the corona are only half of the story about the disturbances that propagate from the Sun to the Earth. The main reason that the connection between terrestrial disturbance and its source at the Sun was so difficult to establish is that both high-speed solar wind streams from coronal holes and coronal mass ejections cause disturbances at Earth, and in both of these types of event the source of the disturbance is not easily detected from ground-based observations, originating as they do from regions that are mostly invisible to such instruments. Moreover, there are other factors that determine whether or not a particular event will be geoeffective – having a strong effect on the Earth – after leaving the Sun, such as whether or not they actually hit the Earth or fly past without coming close enough to hit us. So identifying the source region at the Sun is not enough in itself to establish the connection with events at Earth.

A coronal mass ejection (CME) has two major components: the dense filament and a roughly spherical magnetic front that leads the event; sometimes the cavity between these two components is also counted as a third part of the CME. Illustration 77 shows a large filament eruption that was observed by the AIA coronal imagers on SDO. The successive frames are roughly ten minutes apart, so this portion of the event lasted about an hour.

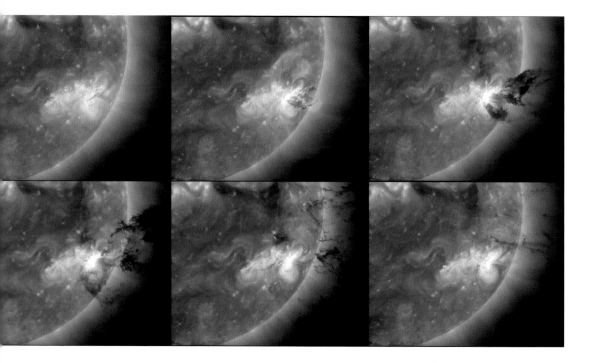

A magnetic front leading a CME, with the filament inside the expanding cavity, is seen clearly in illustration 78, a visible-light coronagraph image from the LASCO instrument on SOHO. As the magnetic front expands outwards it sweeps up coronal material ahead of itself, producing a bubble of bright, expanding emission. The entire eruptive structure of a CME continues to expand as it moves out away from the Sun, as shown in illustration 79.

The expanding magnetic cloud of the CME moves out into interplanetary space, growing in size as it continues on past the inner planets and out past the orbits of Jupiter and Saturn. If the CME is aimed in the correct direction it may hit the Earth on its way out, and if the magnetic field of the CME is properly oriented it may interact strongly with the Earth's magnetic field. Such an interaction is shown in illustration 80, which shows an event coming from the Sun (to the left in this drawing) and hitting the Sunward-facing

77 A large, dense and relatively cool filament is ejected from the solar corona in a major mass ejection, 7 June 2011. Because the filament is so cool and dense, it absorbs the EUV light emitted by the corona behind it, so the filament looks black compared to the bright corona.

78 NRL telescope (a coronagraph) aboard the ESA and NASA Solar and Heliospheric Observatory (SOHO) captured this image of the Sun spewing out a coronal mass ejection (CME) on 28 August 2011. The filament shows up in this white light image as a very bright feature emanating nearly radially outwards from the coronagraph's occulting disc. (The white circle indicates the location of the solar photosphere, which is the source of the light being scattered by the corona and by the dense prominence material.) The leading edge of the CME, a large bubble of magnetic field expanding outwards, is visible above the prominence as an arch of emission.

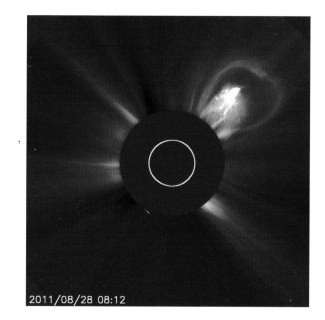

2011/08/28 08:12

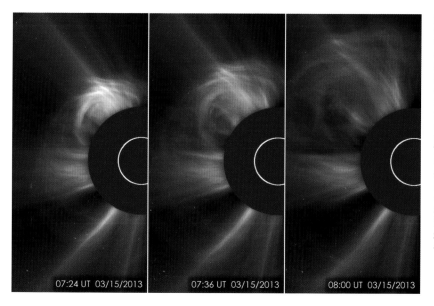

07:24 UT 03/15/2013   07:36 UT 03/15/2013   08:00 UT 03/15/2013

79 NRL's Large-Angle and Spectrometric Coronagraph (LASCO) aboard the ESA and NASA Solar and Heliospheric Observatory (SOHO) captured these images of a fast-moving coronal mass ejection (CME) on 15 March 2013, from 3.24 to 4.00 am EDT. This event reached NASA's Advanced Composition Explorer (ACE) satellite, 1 million miles from the Earth, at 1.28 am EDT on 17 March, and caused a geomagnetic storm a few minutes later.

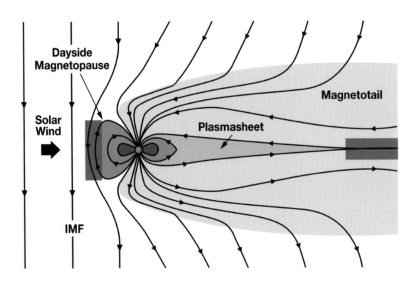

Dayside Magnetopause

Magnetotail

Solar Wind

Plasmasheet

IMF

80 The Earth is surrounded by a protective magnetic shield called the magnetosphere. The solar wind draws the magnetospheric field out into a long magnetotail on the anti-Sunward side of the Earth, and dynamic magnetic events at the forward contact site, the magnetopause, and in the tail can feed back to the Earth, channelled along the magnetic field lines towards the poles. The red boxes indicate the locations where these disturbances begin before propagating towards Earth.

outer extent of the terrestrial field, named 'magnetopause'. In this illustration, the incoming magnetic field is directed oppositely to the terrestrial field and the result is a strong release of magnetic energy (in the region marked by the red shaded rectangle) that accelerates high-energy particles and sends them along the field lines back towards the Earth. Similar reconnection events occur in the magnetotail, the stretched-out magnetic field on the side of the Earth facing away from the incoming solar wind. Here too there is a region where oppositely directed magnetic fields come into contact and reconnect. Those reconnection events accelerate particles, mostly electrons and protons, and feed them back along the magnetic field to come crashing down on the ionosphere of the Earth.

One of the most important functions being carried out in the field of space weather is the forecasting of events that have the potential to be geoeffective. These forecasts are carried out by the u.s. National Oceanographic and Atmospheric Administration (NOAA), the same agency that operates the National Weather Service, and their space weather forecasts have a similar structure,

81 The suite of instruments observing the Sun and the solar wind is used to construct models of solar storms, in order to provide predictions of events that will impact the Earth. See the accompanying text for a detailed description of these figures and what they show.

with alerts and warnings being issued depending upon the predictions from the models of future conditions. Illustration 81 shows one such prediction from NOAA's Space Weather Prediction Center. The sequence of images shows a CME leaving the Sun and indicates that it is expected to hit the Earth about three days later. Each panel shows the density of the ejected material in the top (mostly blue) part and the speed of the material in the bottom

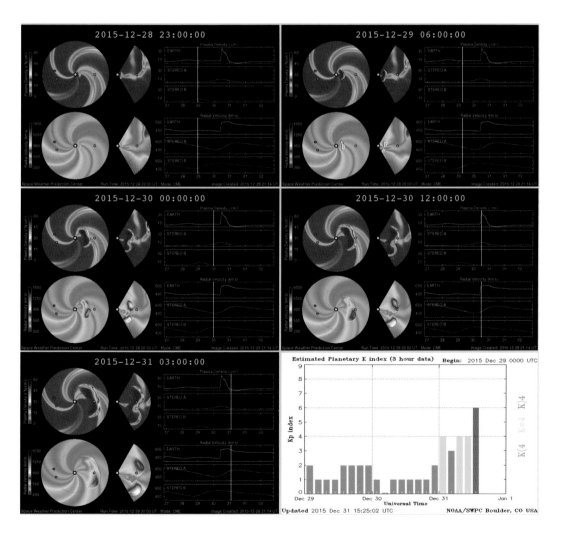

(mostly yellow-green) portion. In addition, the event is viewed from above the plane of the ecliptic (left, circles looking like pinwheels) and along the ecliptic plane (wedge-shaped view). The top left image shows the pre-CME conditions with the pinwheel structures representing the fairly steady outflowing wind, the yellow circle at centre indicating the Sun and the green circle towards the right indicating the Earth, plus the CME just leaving the Sun and heading towards Earth. The top-right and middle-left panels show the CME expanding and evolving as it moves through interplanetary space between the Sun and the Earth. The middle-right image shows the CME hitting the Earth, corresponding to peaks in both the density and speed of the solar wind. The bottom-left panel shows the CME continuing on past the Earth towards the outer planets

82 NASA's Mars Atmosphere and Volatile Evolution (MAVEN) mission has found that solar wind streams are stripping Mars of its atmosphere, and that coronal mass ejections greatly increase that rate when they occur. This rather imaginative illustration shows such an event striking Mars and blasting away atoms of its remaining thin atmosphere. Mars likely had a protective magnetic field in the distant past, but it is no longer present and the atmosphere is now directly exposed to the damaging events from the Sun.

of the solar system. Note that the graphs on the right in each image have a vertical line indicating where in the time sequence each image falls and also indicating the calculated solar wind density and speed for the vicinity of Earth. These five panels were predictions made on the day that the CME left the Sun. For comparison, the bottom-right panel shows the actual magnetic disturbance data at the Earth, indicating the arrival of a magnetic storm late in the day on 30 December, about a half a day later than predicted.

NOAA has a series of geostationary satellites, and in its alphabetically labelled series is up to Geostationary Operational Environmental Satellite (GOES)-R. It was launched in late 2016, carrying a Solar Ultraviolet Imager (SUVI) to observe the Sun with a four-minute series of six ultraviolet filters that are based on the SDO's set of eight filters. (After all, though the GOES satellites' main cameras point towards Earth, the solar panels point towards the Sun, so make a good platform for mounting a solar telescope.) SUVI has a slightly larger field of view than SDO but its resolution is inferior by a factor of 4. A twenty-year lifetime is hoped for.

CMEs continue on past the orbit of the Earth and hit other solar system objects, including any planets that happen to be in their path. The Mars Atmosphere and Volatile Evolution mission (MAVEN) was launched on 18 November 2013, and entered orbit around Mars on 21 September 2014, with the goal of studying the interaction of solar radiation and solar wind with the upper atmosphere of Mars. The available evidence from previous explorations of Mars indicates that the planet had a thick atmosphere in the distant past and was warm enough to support liquid water on its surface. It is likely that Mars also had a protective magnetic field billions of years ago, but due to cooling of its core the magnetic dynamo has ceased and there is essentially no magnetic field now around Mars. The MAVEN measurements show that the solar wind is gradually stripping away the remaining thin atmosphere of Mars and that passing CMEs add to the loss of atmosphere in brief intense bursts during their passage. The loss of atmosphere due to solar events seems to have been a major cause of the changes in Martian climate.

How far from the Sun do these effects extend? The solar wind and coronal mass ejections act together to push against the local interstellar medium – the tenuous material that fills the space between the stars in our neighbourhood – carving out a cavity within which our solar system resides. This volume, the sphere of influence of the Sun, is called the heliosphere and it marks the extent of the solar wind's effects.

Epilogue:
The Heliosphere

O n 25 August 2012, *Voyager* 1 became the first man-made
spacecraft to leave the solar system and enter interstellar
space. Launched from Cape Canaveral on 5 September 1977, its
primary mission was to fly close to Jupiter, Saturn and Saturn's
moon Titan.[24] The spacecraft reached Jupiter on 5 March 1979,
recording detailed images of the planet and several of its moons,
most notably detecting active volcanoes on Io. With the help of a
gravity-assist manoeuvre, it went on towards Saturn, with closest
approach on 12 November 1980. It then continued outwards, and
is presently about 20 billion km from the Sun.

We've seen that the solar influence extends out past the Earth, past
the large gaseous planets Jupiter and Saturn, farther out than all of the
known solar system planets. How far out does the sphere of influence,
to use a term borrowed from politics, extend? Is there a 'heliosphere'?
Do the emanations from the Sun gradually taper off without a clear
end point, or is there some sort of clear boundary between our solar
system and interplanetary space? One might expect the solar wind to
behave the way light does, gradually spreading out in all directions
from the Sun, diminishing in intensity as it spreads out over a larger
and larger area. Viewed from far away, the Sun grows smaller and
fainter as we move farther away from it, becoming a small point of
light like any other star, and eventually becoming too faint to see by
eye. The changes happen smoothly and without any obvious boundary

or discontinuity. The same holds for the gravitational pull of the Sun: it extends outwards in a smoothly diminishing way, gradually becoming weaker, in principle extending out to infinity but in practice becoming negligible at a distance of about 3.6 light years.

But the solar wind behaves differently. There is an interstellar medium, a gas of very low density between the stars, more than a billion billion times less dense than our terrestrial atmosphere. Near the Sun, the outflowing solar wind is strong enough to push back this background gas, so it creates a large volume of space that is filled with the expanding solar coronal material and the additional ejections of hot magnetized plasma that erupt from time to time. Eventually though, the strength of this wind becomes too feeble to push the interstellar material out of the way, and there turns out to be a boundary beyond which we can say that the influence of the wind has stopped. We name this boundary the heliopause, where 'pause' simply means 'end'.

The outer boundary of the solar wind's influence was predicted by Eugene Parker when he formulated his theory of the solar wind and noted that it would expand at supersonic speed through inter-planetary space. Moving farther and farther away from the Sun, the increasingly feeble wind is eventually slowed down to the point that it is no longer supersonic. The transition happens via an effect that is peculiar to supersonic travel. The key principle in any given medium is that disturbances will propagate at a particular speed that is determined by the medium's ability to recover from the disturbance. If you compress air, or a spring, or anything similarly elastic, it will spring back, and that reaction will happen at a speed that depends on the nature of the medium that is disturbed. A soft material that has a great deal of give will react slowly, so the disturbance propagates slowly. A very stiff material, such as a metal bar, has very little give and will react quickly, so the disturbance propagates at very high speed. The name we give this speed is 'sound speed', though it clearly applies to more than just sound

83 Tail of Comet McNaught seen from STEREO, 18 January 2007. The overall shape of the tail is produced by dust being left behind as the comet moves in its orbit around the Sun. The striations, however, are due to the interaction of the ionized particles in the comet tail with the magnetized outflowing solar wind, and they point away from the direction of the Sun, which is to the right out of the frame of this image. The bright star-like features with vertical lines left of centre and at the right edge of the image are, respectively, Venus and Mercury. They are so bright that their light saturates the sensitive camera used to record this image.

waves, such as to the more general type of wave produced by compressing an elastic medium.

What happens when an object creates a disturbance by passing through a medium? An aeroplane does this, for example, pushing air out of its way as it flies through. All is well as long as the plane is flying slowly enough for the air to be able to move away. The disturbance created by the plane moves away at the sound speed – in this case literally the speed of sound – and races off ahead of the plane. But if the plane travels faster than the speed of sound, the air is unable to get out of the way quickly enough. The plane compresses the air and piles it up in a wall of higher density. This wall, called a 'shock front', propagates away to the side of and behind the plane at the speed of sound. When it passes us on the ground, we hear this compressed front passing us as a sonic boom.

A related phenomenon happens with the supersonic solar wind expanding out into interstellar space. The expanding supersonic solar wind eventually makes a transition to subsonic speed and we get the reverse situation: the subsonic wind impedes the faster supersonic wind and there is again a high-density wall formed. The transition between the two is a shock wave, in this case a standing shock, one that remains more or less in the same place at all times. We say 'more or less' because the expanding solar wind is variable, so the transition can occur closer in or farther out, depending on its outflow speed. Still, all around the Sun there is an abrupt, clearly defined transition called the 'termination shock' that we can call the extent of the influence of the solar wind. Both of the Voyager spacecraft have been seen to cross this termination shock. *Voyager* 1 is believed to have crossed it in December 2004, at a distance from the Sun of 94 au, while *Voyager* 2 seems to have crossed in May 2006, at a distance of 76 au.

The Heliopause

The termination shock is only one of the boundaries defining the limit of the Sun's influence. The solar wind may have slowed down, but it is still flowing outwards and its pressure is still large enough to push back the interstellar gas. After the termination shock, the solar wind interacts with the interstellar gas in a swirling, turbulent region given the name 'heliosheath'. Finally there is another region, farther out, where the outflowing solar wind is no longer strong enough to press back against the interstellar medium and it is brought to rest. Beyond this heliopause, we can truly say that we are out of the solar system and in interstellar space.

In one of the most stunning achievements in the history of spaceflight, *Voyager* 1 seems to have passed through the predicted turbulent region for several years and then to have crossed the final outer boundary as of 25 August 2012 (illus. 84), as indicated

84 This plot shows the number of cosmic ray particles per second detected by *Voyager* 1 from October 2011 through October 2012. The detected rate begins to fluctuate starting in August 2012, finally settling in to a sharply lowered number of solar cosmic ray particles by September of that year. This transition is one way to mark the crossing of the outer extent of the Sun's influence.

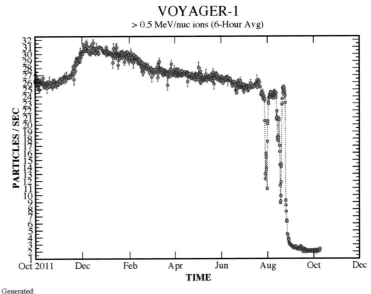

VOYAGER-1

> 0.5 MeV/nuc ions (6-Hour Avg)

Generated:
Wed Oct 10 14:16:24 2012

by a sudden drop in the number of cosmic rays. These high-energy particles are no longer coming from the Sun, but appear to be from the region beyond the solar system. Instead of high-energy particles coming from the Sun – solar cosmic rays – we are now seeing a different class of high-energy particles that permeate interstellar space: galactic cosmic rays.

Ploughing through the Interstellar Medium

The Sun, along with its magnetic field, its outflowing solar wind, and all of its planets and countless other smaller objects, is moving through a cloud of gas and dust known as the Interstellar Medium (ISM). The outflowing solar wind eventually encounters this surrounding medium, and there should, in principle, be a transition between the two. To get some idea of what our solar system might look like when viewed from another distant star, we can look at other stars moving through their own interstellar surroundings. One of

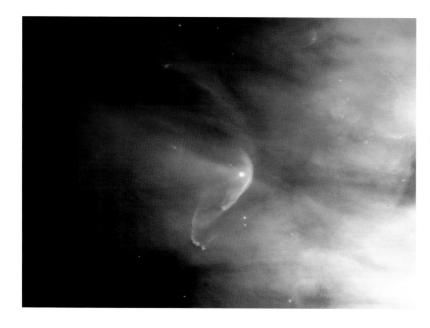

85 This arcing, graceful structure is actually a bow shock about half a light-year across, created as the wind from young star LL Orionis collides with the Orion Nebula flow. Part of a mosaic covering the Great Nebula in Orion, this composite colour image was recorded in 1995 by the Hubble Space Telescope.

the best known of these regions is located in the constellation Orion, where we see clearly the interaction between young stars and the local ISM as they plough through it (illus. 85). There is some question now about whether or not the wind from the Sun produces an impressive arc-shaped structure known as a bow shock, but it is certainly the case that we see these structures forming around other stars, especially in the direction of their motion relative to their local interstellar medium. To illustrate our place in the galaxy, we end with this image of a star in the Orion Nebula, LL Orionis, taken with the Hubble telescope. Adrift in the nebula, the young star LL Orionis produces an energetic stellar wind that runs into the slower sur-rounding gas and forms a shock. Another small bow shock can be seen ahead of a faint star at the upper right. We are most certainly in a similar position, far from our nearest neighbours, drifting through swirling interstellar plasma, living on a small sphere of rock that accompanies the glowing ball that we call our Sun.

APPENDIX 1:

OBSERVING THE SUN SAFELY

The Sun is almost a million times brighter than the full Moon, and is much too bright to look at safely with your naked eyes. In seconds, you could cause permanent damage to your eyes if you stare at it. And certainly don't use binoculars or a telescope to look at the sun directly, unless they are properly filtered. The optical devices' concentration of unfiltered sunlight can cause a retinal burn quickly.

The eye (the retina, in particular) doesn't have pain sensors, so you wouldn't know from pain that you were damaging your eyes. Imaging the Sun on your retina could make a retinal burn, damaging or destroying the rods and cones (see Chapter Four). Sometimes ophthalmologists detect a crescent-shaped retinal burn that matches the shape of the eclipse crescent that someone had observed.

Sunglasses don't help enough when looking directly at the sun; they cut the solar intensity by a factor of only 2 or so. What you really need are the increasingly available 'eclipse glasses', more accurately 'partial-eclipse glasses', or really 'full-sun-watching glasses'. They only cost a pound or a dollar and less in bulk. They use a dark polymer (dark because of suspended carbon particles) or an aluminized mylar (not, technically, the trademarked Mylar) to absorb all but one part in about a hundred thousand of the incoming sunlight. There is an International Standards Organization (ISO) approval number that could appear, and you could choose to accept only filters that show that standard certification, which comes along with a big CE logo;

the expected certification as of 2016 is ISO 12312-2. If you hold the glasses up to the sun far from your eye, you can see if there are any pinholes of any size in them; reject glasses if you see bright points of light coming through them (illus. 86).

Once upon a time, people used smoked glass to observe the sun; basically, you coat a glass plate with soot from a flame. But such coatings are uneven and can easily be wiped off irregularly, so we now consider them to be unsafe. Similarly, looking through the coatings on DVDs or CDs can cut down incoming sunlight to an acceptable level, but non-experts may not appreciate when the absorption/reflection of incoming sunlight is sufficient – and novices may even look through the hole, which gives completely unfiltered sunlight.

Another typical piece of advice is to use number 14 welder's glass. Number 12 will actually do, though it gives a brighter solar image, and is more readily available. But the inexpensive eclipse glasses are so easily available and cheap that it makes more sense to use them. And the welder's glass gives a greenish image, which some find less pleasing than the orangish image that results from several other filter types.

Traditionally, fogged and exposed black-and-white photographic film, or x-ray film (look through the dense parts, not the transparent parts caused by the bones!) absorbs enough light to make a filter, but these days old-fashioned black-and-white film is rare. Colour film should not be used as it doesn't have the absorption in the infrared that made black-and-white film safe.

We use the term 'neutral density' to indicate a logarithmic scale of attenuation across a wide range of the spectrum, though neutral density in the visible part of the spectrum doesn't mean that there isn't an 'infrared leak' that can pass enough invisible radiation to hurt the eye. ND0 is no attenuation, that is, 100 per cent transmission; ND1 is one power of ten attenuation, yielding 10 per cent transmission; ND2 is 1 per cent transmission. For safe solar viewing, the main expert on solar filters, the optometry professor emeritus B. Ralph Chou of the University of Waterloo, Canada, recommends density of at least 4.5 in visible light and about 2.3 in the infrared; that is, transmission of less than 0.003 per cent in the visible and 0.5 per cent in the infrared. His work (see URL below) includes a page of graphs of various filters' transmission as a function of wavelength.

For use with a telescope, many people prefer to use neutral-density filters made of a glass substrate with a chromium deposit. For telescopes (or binoculars), filters should always be used at the front of the instrument so that the sunlight is filtered before it enters the optical elements. That way, the full sunlight isn't focused onto the filter, which could be a hazard and even crack a filter or an optical element. One can buy full-aperture neutral-density filters.

Some advanced amateurs use special filters that pass the red hydrogen spectral line or the ultraviolet ionized-calcium spectral line. These show structure in the solar chromosphere. Neutral density filters, on the other hand, show the photospheric layer of the solar atmosphere, including the sunspots.

With a little advance work, you can order solar-eclipse viewing glasses from various providers, especially:

Europe: Assistpoint, Ltd, www.eclipseglasses.co.uk
America: Thousand Oaks Optical, www.thousandoaksoptical.com
Rainbow Symphony, www.rainbowsymphony.com
American Paper Optics, www.eclipseglasses.com (but don't confuse red/green 3-D glasses with the full-absorption solar glasses).
Baader Planetarium (Germany), www.baader-planetarium.com/.
For safety and ISO (film speed) information, see www.eclipseglasses.com/pages/safety.

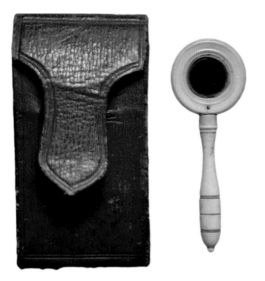

87 Ivory eclipse-glass holder from the 1790 solar eclipse in London. Just what filter material was used is not clear.

See an online exhibition of solar glasses at: http://astronomy. williams.edu/hopkins-observatory/eclipse-viewers, including an example dating back to the solar eclipse in London in 1790 (illus. 87). A selection of the actual filters is on display in the Milham Museum of Astronomy in the Hopkins Observatory of Williams College in Williamstown, Massachusetts.

For a school class, you might prefer to buy a sheet perhaps 50 cm or so square, and put it in a cardboard frame, so that young students can easily stand behind it (illus. 88).

See also the section on *Eye Safety and Solar Filters* at www.eclipses. info, the website of the Working Group on Solar Eclipses of the International Astronomical Union.

http://uwaterloo.ca/optometry-vision-science/
    people-profiles/b-ralph-chou
www.skyandtelescope.com/observing/solar-filter-safety

88 Safe eclipse viewing, using individual glasses (*top*), or (b) a large sheet of filter material (*bottom*).

APPENDIX II:

AMATEUR SOLAR OBSERVING

To see sunspots safely, you need to use a telescope that is suitably filtered at its front end ('solar filters' at the back end can break under the intense focused solar energy, so aren't considered safe). 'Neutral-density' filters to fit on the front of any telescope are available for a price of order $/£50. That is, they cut down all colours of light fairly equally, to give a white image – or sometimes pass a little more orange to make a pleasing orange image. A compact design with the optical path folded with mirrors to make the device small and portable is known as a 'Sunspotter', and is available from sources below for a few hundred pounds or dollars. Also consider Solarscope, both a very inexpensive one made of cardboard and a newer version made of wood, made in France and available at sites in the U.S., www.solarscope.org/us/index.us.html, and UK, www.solarscope.org/en/index.en.html.

To view the reddish hydrogen-alpha line of the solar chromosphere and prominences that stick above the solar limb, telescopes are available with filters installed for about £1,000. Such filters could also be installed on existing small telescopes.

A brief history of sunspot observing is at: http://solar-center.stanford.edu/about/sunspots.html, with discussion of the discovery of sunspots at www.nasa.gov/mission_pages/sunearth/news/400yrs-spots.html. The following groups of amateur astronomers are devoted to studying the Sun:

A subgroup of the American Association of Variable Star Observers (AAVSO) is dedicated to observing sunspots: www.aavso.org/solar.

The Solar Section of the Association of Lunar and Planetary Observers (ALPO) posts observations at: alpo-astronomy.org/solar. The Charlie Bates Solar Astronomy Project is devoted to solar outreach: www.facebook.com/groups/charliebatessolarastronomyproject/. Among many Facebook groups one devoted specifically to solar observing is www.facebook.com/groups/solaractivity/.

You can order neutral-density solar filters, for sunspot observing, mounted to fit your telescope from various providers, especially: Thousand Oaks Optical, http://thousandoaksoptical.com.

You can order hydrogen-alpha filters, to view the solar chromosphere and prominences (or, for more advanced viewing, filters for viewing at the wavelengths of the calcium H or K lines), for your telescope from:

Daystar Filter Corp, http://daystarfilters.com
Lunt Solar Systems, http://luntsolarsystems.com
Solarscope (Isle of Man), www.solarscope.co.uk.

You can order solar telescopes with suitable H-alpha or calcium-line filters fitted from:

Daystar Filter Corp, http://daystarfilters.com.
Coronado 'Personal Solar Telescope' ('PST') with an H-alpha or K-line filters from various camera stores and telescope suppliers; consult the manufacturer's website at: www.meade.com/products/coronado.html.
Lunt Solar Telescopes, http://luntsolarsystems.com
Solarscope (Isle of Man), www.solarscope.co.uk

A crowd-sourced site for categorizing and ranking sunspots is at www.sunspotter.org.

A compact device, a 'Sunspotter', for viewing a solar-disc white-light image to see sunspots is available at www.scientificsonline.com/product/sunspotter and www.teachersource.com/product/sunspotter-solar-telescope/astronomy-space.

APPENDIX III:

OBSERVING THE CORONA AT ECLIPSE

The most glorious sight anyone can see, in our opinion, is a total solar eclipse. To have the sky go dark in the middle of the day, with fantastic phenomena occurring overhead, is a fabulous experience. But total solar eclipses occur only about every eighteen months somewhere in the world, and then only in a path perhaps 100 km wide and thousands of kilometres long. You must be in this path to experience totality.

Seeing the total part of the eclipse can be inspirational for students and others. Therefore, everyone should be encouraged to travel into the zone of totality. Amateur-astronomy societies or planetariums usually have public viewing locations. Teachers can help their students to obtain, and observe through, proper filters (illus. 89).

Even having the Sun eclipsed 99 per cent won't do. The corona is about a million times fainter than the bright disc of the Sun, so even if only 1 per cent of the solar photosphere is visible, what is left is still 10,000 times brighter than the corona. As we described in the previous appendices, whenever even a small part of the every-day Sun (the solar photosphere) is visible, you need to use special solar filters in order to look at the Sun safely.

Even if you are in the zone of totality, you need partial-eclipse glasses (often simply and a bit misleadingly called 'eclipse glasses') for the hour or so of partial eclipse leading up to totality and for the

89 Class of four-year-olds with both a foot-square (half-metre-square) large filter in a cardboard frame and individual 'partial-eclipse glasses' (one child testing them by looking in the wrong direction, so she saw nothing).

similar amount of time after totality. Only in the last few minutes of the partial phases before totality would you even notice that an eclipse was occurring, unless you were able to look up through special glasses.

Indeed, we prefer filter cards – hand-sized cards with filter material in them that you have to hold up instead of having earpieces that allow you to wear them. The partial phases change slowly, and it is enough to glance up for a few seconds through a filter every five minutes or so. So better to avoid the temptation of staring at the Sun for minutes on end, even though it is safe to do so if you are properly wearing the proper partial-eclipse glasses.

When (through the eclipse glasses) the solar crescent begins to shrink from point to point, you are almost at totality. When the crescent shrinks completely as seen through the glasses, then you are at the time of Baily's beads; starting then, you can look at the eclipse directly, though don't stare long while the bright white beads are there – just glance for a second or two. The last Baily's bead is the diamond ring (rarely a double diamond ring), and in the few seconds that the diamond ring is diminishing, it becomes safe to look at the eclipsed Sun directly.

90 This wide-angle image was taken during the total solar eclipse of 9 March 2016 from the Indonesian island of Ternate, with all the eclipse's totality phenomena visible though through thin cloud. Taken with a Nikon D600 with Nikkor 24–85 mm f/8 zoom lens at 24 mm. Partial-phase and totality images are shown across the top, the former taken with a Nikon D7100 with Nikkor 80–400 mm lens at 400 mm and Questar filter. Totality image used a Nikon D7100 with Nikkor 500 mm f/8 lens, no filter.

Traditionally, one can see the crescent of the solar photosphere (the surface of the Sun that we see every day) with a 'pinhole camera', just making a hole a few millimetres across in a piece of cardboard and looking down at what it projects, with the sun safely at your back. But partial-eclipse glasses are so readily available and inexpensive now that pinhole-camera images seem dim and unexciting.

A person knowledgeable about observing the Sun could also project an image with a telescope or binoculars (held backwards), but again, keep the Sun at your back and look only at the ground, wall or screen on which the image is being projected.

As soon as the diamond ring disappears, and you take down your filter from in front of your eyes, you will see the reddish chromosphere and the pearly white corona surrounding it and the black silhouette of the moon. The corona is visible for the duration of totality – two minutes to two minutes forty seconds or so on totality's centreline for the American eclipse of 2017, and up over six minutes for the longest possible totality these days as viewed from the ground.

Only during this period of totality is it safe to look at the Sun directly. Contrary to what some people mistakenly think, there are no additional rays that come from the Sun during an eclipse; we merely see the corona that has always been there behind the blue sky, but the blue sky is taken away.

No change can be seen within the minutes (or seconds) of totality from one spot, but the Moon's shadow takes two or more hours to traverse its long path across the face of the Earth, and comparing images from different locations along the path shows changes in coronal streamers and polar plumes.

91 Modern set of partial-eclipse glasses from the Luke Cole eclipse viewer collection. The glasses need to be removed for you to see totality once the bright photosphere is completely covered. Occasionally, people aren't given correct instructions about these misleadingly entitled 'eclipse glasses' (they really should be named 'partial-eclipse glasses') and leave them on during totality – so they see nothing of totality because the solar corona is much too faint to be seen through these solar filters.

If you are a first-time viewer of totality, you are usually advised to just look, and not divert your attention from the spectacle by taking photographs, possibly forfeiting the full experience. After all, more experienced astronomical photographers will take lots of photos that you can access (illus. 90). But if you can't resist photographing the eclipse, then you have to determine how serious (how professional) you want your equipment to be. At the Svalbard eclipse of 2015, great photos and movies were taken with iPhones. But often if you are using just a phone camera or a pocket camera, the autofocus may 'hunt' throughout the minutes of totality, never reaching a focus.

92 Both sides of a rectangular 'card' with filter material; it can just be held up in front of your eyes with one of your hands, which reminds you merely to glance for a few seconds every few minutes during the partial phases. It is not to be used during totality.

Sometimes instead of taking telephoto images in which the eclipsed Sun is large in the frame, it is more interesting or more fun (even if less scientific) to take a wide-angle view, showing the setting of the totally eclipsed sun in the landscape. Only during totality do such attempts work well, since even when only a little of the uneclipsed crescent is in the photographic frame, it will be overexposed and, unless there happens to be the right amount of haze or clouds, its shape will be hidden so you can't see in the image that an eclipse is occurring.

One of us (JMP) is involved in a Megamovie project, based at Lawrence Berkeley Laboratory, in which crowd-sourced eclipse photos of the 2017 totality will be uploaded (courtesy of a Google-company arrangement) and assembled into a whole movie. We may wind up with separate movies for images taken with smart-phones and images taken with ordinary cameras. See http://eclipsemegamovie.org.

The Citizen Continental America Telescopic Eclipse (CATE) Experiment, organized by Matt Penn of the National Solar Observatory and others, plans to have observers with identical equipment spaced at dozens of locations across the United States in 2017, after a trial at the 2016 Indonesian eclipse. See https://sites.google.com/site/citizencateexperiment/home/.

If you are using a standard single-lens-reflex (SLR) camera (Nikon and Canon are common brands), you will want to use a substantial telephoto lens. Traditionally, a 500-mm focal length lens is ideal for a full-frame camera; many cameras now use 2/3-size chips (that is, chips 2/3 the size of full-size chips – FX in Nikon's terminology, which match the size of a frame of 35-mm film – which few people use anymore, given the electronic revolution). On a 2/3-size chip (DX, in Nikon's terminology), a 300-mm lens is roughly equivalent. Such telephotos give the full solar disc

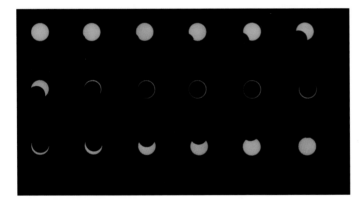

93 A series of images at the total eclipse of 2009 from China, first partial phases through a partial-eclipse filter, then totality with no filter and then the final partial phases through the partial-eclipse filter again.

94 A series of images at the annular eclipse of 2016 from La Réunion in the Indian Ocean; since the Sun was not totally eclipsed at any time, the partial-eclipse filter had to be used throughout.

95 A series of images at the partial solar eclipse of 2015 from South Africa; since the Sun was not totally eclipsed at any time, the partial-eclipse filter had to be used throughout, except for the brief times in the early morning when cloud cover was so dense that the partially eclipsed Sun could be viewed through the clouds.

with the corona around it extending outwards perhaps for a solar radius or so, and some aesthetically interesting sky around the corona (also providing for some motion for the Sun during the minutes of totality).

It is very important for all telephoto eclipse photography to mount your camera on a sturdy tripod. Also, try to get a cable (or wireless control) that releases the camera's shutter without your having to touch the camera – which would introduce some shake.

Normally, during the eclipse, you would take a graded series of exposures, so-called 'bracketing'. So you might take 1 s, 1/2 s, 1/4 s, 1/8 s, 1/16 s, 1/32 s. It is best to set the 'speed' (the ISO setting) to some moderate level such as ISO 800; though cameras may have settings with ISOs 8 or more times that, the higher settings usually make noisier (grainier) images. Set the lens opening (if it is adjustable; some long telephotos are mirror lenses and are not adjustable) to one or two f/stops (f is the focal ratio, the lens's focal length divided by its diameter) more closed than wide open, since often lenses are not quite as sharp at their widest setting. So if your camera lens has a range from f/4 to f/32, set it for one stop higher than f/4, which is f/5.6 (if you square both numbers, you will see that the area of the opening is a factor of 2 different between those two settings). One nice thing about photographing an eclipse is that since there is such a wide range of brightness in the corona – about a thousand times brighter at the solar limb than at one solar radius farther out – every exposure is good for some portion of the corona. A bracketed series of exposures will show a range of coronal extent.

Don't forget to download and to back up (make a copy of) your eclipse images; you don't want them to disappear by mistake!

After the minutes of totality, all of a sudden the diamond ring will begin to brighten on the other side of the solar disc from the diamond ring at the beginning of totality. That's the sign for you to look away and to again start observing the eclipse only through one of the special partial-eclipse filters (illus. 91) or filter cards (illus. 92).

APPENDIX IV:

OBSERVING THE SUN FROM SPACE

Carrying out observations from space is a difficult business. Suppose that you have a telescope and want to point it at a target and take some photos. How would you go about it in space? You can't set it on the ground, because there isn't any ground. You can't look through the eyepiece because you are probably down on Earth, or if you're an astronaut you likely are inside a protected environment while the telescope is outside. For similar reasons, you can't point the telescope at your target, and if you do take a picture then what do you do with it? Going back a few steps, how did your telescope get up into space? It is presumably a high-quality optical instrument, so how did it survive a punishing launch, filled with vibrations and accelerations? Once out in space it's in a vacuum, with the Sun heating one side of it up to hundreds of degrees (there is no air to cool it) and the other side radiating heat so it cools down to well below freezing. There will thus be expansion and contraction of the respectively hot and cool parts of the telescope, leading to twisting and distortion of the optics and associated structures, thereby causing severe focus and aberration problems, while high-energy particles penetrating the system will cause damage to any electronics.

To set up an observing site in space, there are four major components needed:

1   The instrument that you intend to use (illus. 96), suitably designed to operate in the harsh environment of space, including electronics to convert the image or other desired signal into a form suitable for radio transmission down to the ground.

2   A support system generally called a spacecraft, which supplies power, pointing (to steer the instrument towards the intended target), data processing, data storage and telemetry.

3   A launch vehicle, to bring the observatory (instrument plus spacecraft) into space.

4   A ground station, to receive the data and make it available to the scientists on the ground.

We'll go through this list item by item and talk about the special needs and requirements of observing from space. First, though, we ask why one would want to carry out astronomical observations from space, since it is a far more difficult business than observing from the ground. The short answer is that we observe from space when it is the only way, or at least the best way, to get the desired data. For instance, wavelengths of light in the ultraviolet do not reach the ground because they are absorbed in the Earth's atmosphere, while many of the strongest and most important wavelengths from the Sun and stars are in the UV. The only way to observe those wavelengths is to put the observing instrument above the atmosphere.

The Instrument

The single most important consideration about putting an instrument in space is to understand, deeply and fully, that once it has been launched you will never again be able to touch your instrument. No tinkering with it, no rewiring, no fixing of a broken part – nothing can be done if it involves a person physically handling it. (Manned space flight is the exception, but that accounts for only a small fraction of the instruments in space.)

96 Telescope assembly, wrapped in insulating foil and integrated onto the spacecraft. It is not very beautiful, except perhaps to the eyes of the scientists and engineers who built it.

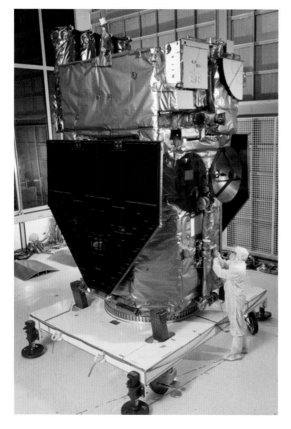

97 When the instruments are integrated onto the spacecraft, the unit as a whole is designated an 'observatory'. The four telescopes comprising the Atmospheric Imaging Assembly (AIA) are sticking up at top left, on the far left side of the support structure of the Solar Dynamics Observatory.

From the very first stages of the design process, this basic fact must be central to one's thinking.

The next most important factor is to understand the environmental conditions that the instrument will experience in space. The side of the instrument facing the Sun will get extremely hot, typically above the boiling point of water unless special steps are taken, and there is no air to cool it or to redistribute the heat. So while one side may get hot, the other side will usually be looking out into deep space and will radiate away its warmth until it gets to temperatures as low as −100°C, again unless special steps are taken. Such huge temperature gradients could deform and even incapacitate a delicate, precision instrument, which is why they are often wrapped in a thin, lightweight reflective and insulating material. If necessary, heaters are glued to the instrument body under this insulating foil, to maintain the desired temperature.

Next on the list of dangers is surviving the launch. The vehicle used to put the instrument in orbit produces violent acceleration and vibrations, enough to tear apart most normal pieces of equipment. The sophisticated, delicate observing tool that you build must survive being shot out of a cannon (metaphorically) and then survive bone-rattling vibrations that may last for many minutes. A host of special design techniques is employed to ensure that the transfer of vibrations from the rocket motor to the instrument is minimized and that the instrument then can survive the remaining vibrational forces. Extensive testing on the ground is carried out to ensure that the instrument survives, and redesign is often necessary after a test if the results aren't satisfactory.

Implicit in the design from the outset is concern about radiation. Space is filled with high-energy particles called cosmic rays, both from the Sun and from interstellar space. These particles penetrate the instrument and can interfere with its electronics, causing errors or even destroying crucial components. Shielding crucial places, such as the sensors, can be done, but shielding is heavy and

98 The Solar Dynamics
Observatory (SDO)
was launched into
geosynchronous orbit on
an Atlas v rocket.

the mass allocation for an instrument is usually strictly limited. Specially designed versions of electronic components less sensitive to radiation damage must be used. These take time to develop and test, so that the onboard electronics are typically out of date by ground-based standards.

The Spacecraft

The observing instrument is typically mounted to a life-support system known as a spacecraft, which supplies power, pointing capability (if that is needed), computational resources and radio contact so that data can be sent to the ground and often so that commands from the ground can be sent up to the instrument (illus. 97). Building spacecraft is a specialized skill and experienced teams of engineers are needed for this. For scientific research, spacecraft are often built by private aerospace companies under contract to NASA.

99 The SDO downlinks science data from onboard antennas to the ground station, consisting of two 18-m antennas located at White Sands Missile Range, New Mexico.

The process of tying together, or integrating, the instrument to the spacecraft can be long and involved, for reasons similar to those described above for testing the instrument itself. The entire object, now called an observatory, needs to be tested to ensure that all of its functions work as expected, that it survives the launch vibrations and that it behaves thermally as expected once in orbit. These environmental tests can sometimes go on for many weeks before the observatory is declared ready for launch.

The Launch Vehicle

Typically, the single most expensive item in the entire process of building an instrument and putting it in space is the launch. The basic issue here is that the payload – the object that is to be put into orbit – needs to be propelled up to extremely high speeds. To orbit the Earth, a satellite must reach a speed of about 5 miles per second, or else it will just fall back down to the ground. To escape the Earth's gravitational pull, it must reach a speed of 7 miles per second. Either way, a huge amount of energy needs to be expended to accelerate anything to such high speeds. The best method we have at this time for doing that is to burn fuel violently in a combustion chamber and expel it at very high speed from the rear nozzle of a rocket. Despite many decades of effort, no better method has yet been found.

A moment's thought makes clear that the less mass you bring up into space, the less fuel you will need. This is why orbital rockets are almost always multi-stage: the first stage starts the acceleration process going, and when the fuel is used up the large fuel tank is discarded. The second stage then takes over, without having to accelerate the huge mass of the first stage's empty tank. Some launch vehicles dispense with the first stage by being carried to high altitudes under a large jet travelling at hundreds of miles per hour, and then dropped for launch. This method can be used for small and medium-sized payloads (illus. 98).

The Ground Station

Most scientific missions require that data be brought back down to the ground, to be analysed by the science team. Typically that task is accomplished via radio contact, using a transmitter and antenna in the spacecraft, and a large antenna coupled to a receiver on the ground (illus. 99). The amount of data that needs to be sent down will strongly affect the cost of operating the observatory, because these data downlinks can be rather expensive. Depending on the orbit, the ground contacts might be limited to just a couple per day in equatorial orbit, to about sixteen contacts per day in polar orbit, to nearly continuous contact in geosynchronous orbit. Once the data reach the ground, they must be relayed to the scientists, typically over the Internet. For an imaging mission with large images and continuous viewing, special high-speed data transmission lines may be necessary.

ENDNOTES

1 An earlier version of this chapter appeared in the book *Making Sense: Beauty, Creativity and Healing*, ed. Bandy Lee, Nancy Olson and Thomas P. Duffy (New York, 2015).

2 Herschel's discovery came about in a wonderful way: Herschel was attempting to measure the amount of power in sunlight at different wavelengths. He did this by passing a beam of sunlight through a prism and splitting it into its component colours. He then moved a thermometer across the spectrum to measure how much it warmed and thereby to measure the amount of energy coming from the Sun at each wavelength. He found that the red light from the Sun warmed the thermometer even more than did the yellow and blue wavelengths, which he interpreted as showing that red light had a greater warming power. What he actually had discovered is that more of the Sun's energy is emitted in the red light than in the blue. His report of this observation was not believed, so he refined the experiment by adding a control thermometer outside of the beam of light coming through the prism to rule out an overall warming of the room itself. He then also made measurements out of the visible region, in a location beyond the red where there was nothing visible to the eye. He found that the thermometer warmed up there as well, even more than in the red light, and after repeated measurements he concluded that there was a colour out there beyond the red end, invisible to us but containing a substantial amount of energy. This region of colours beyond the red is now called the infrared.

3 As a young researcher in the 1970s, one of us (LG) was told that it would be difficult to do anything new that Hale had not already done sixty years earlier. He thought the comment was meant as a joke, but it turned out to be the plain truth.

4 Some plastic solids, such as taffy, can also be considered fluids. This gets a bit tricky though, as the speed of the deformation becomes a factor. Silly

Putty will deform slowly under gravity, but will bounce like a solid when pushed quickly (hitting the floor, for instance).

5   What produces a wave? Two things are required: a disturbance and a restoring force. A piano string, for instance, is struck by the hammer moved by one of the keys, which displaces the string from its rest position. But the string is under tension, so there is a force that pulls the string back to its rest position. A wave then propagates along the string from the spot that was struck by the hammer, and soon the entire string is moving back forth all along its length. Without the hammer's disturbance there would be no wave produced, but also without the restoring force, the disturbance would merely displace the string and nothing would vibrate after that. The wave that propagates as a result of the disturbance moves with a certain speed, generically called the sound speed, which depends on two quantities: the strength of the restoring force and the mass of the string. A very heavy, thick string that is under a weak tension will vibrate slowly; tightening the string increases the restoring force and changes the ratio of the mass to restoring force. This causes the sound speed to increase, which in turn makes the string vibrate more quickly, so the frequency of the oscillation increases.

6   Scientists sometimes trace their lineage back through a chain of thesis advisors. Via Leighton's PhD advisor, William Vermillion Houston, Leighton was academically descended from Houston's PhD advisors, Robert Millikan and Albert Michelson, who were both Nobel Prize winners.

7   Imagine the wave propagating as a series of crests, like ocean waves hitting a beach. The direction of the wavefront crest is perpendicular to the direction of propagation. The series of crests move forward and encounter, at an angle, a boundary line where the crests suddenly progress more slowly. When the first wave crest reaches the line, the part of it that hits the line slows down, while the rest of the crest keeps marching forwards. The result is a bend in the row, and ultimately a bending of the entire line. It's like driving a car that has brakes only on the right side. When you step on the brake pedal the right side slows down, but the left keeps going. The result is that the car turns towards the right.

8   There is something of a puzzle related to the standard version of the story, claiming that Schwabe was searching for Vulcan. He began his sunspot observations in 1826, but it wasn't until 1840 that the mathematician and philosopher François Arago convinced his colleague Urbain Le Verrier to begin working on the orbital motion of Mercury, and it wasn't until many years later, in 1859, that he proposed an inner planet named Vulcan in order to explain the mismatch between his calculations and the observed motion of Mercury. So the explanation that is usually given for Schwabe's studies doesn't match the timing, being off by three decades. In fact, the Royal Society citation written by a Mr Johnson in 1857 is more careful: 'I am not

aware of the motive that induced him – whether any particular views had suggested themselves to his own mind – or whether it was a general desire of investigating, more thoroughly than his predecessors had done, the laws of a remarkable phenomenon, which it had long been the fashion to neglect.' No mention of Vulcan or inner planets at all!

9 Crucially, the spot fields are not precisely parallel to the equator, but instead tend to be tilted a bit, with the leading spot closer to the equator than the trailing spot. Higher-latitude spots are more tilted than are low-latitude spots; this correlation is known as 'Joy's law'.

10 The way that more rapid rotation of the equator stretches out the magnetic fields that have emerged at the surface can also be seen in the sequence of images shown in illus. 72, where vertically oriented features such as coronal holes become V-shaped after several months due to the differential rotation.

11 For those who champion Sydney Chapman, I note that well into the late 1940s he was promoting a view of terrestrial and solar magnetism based on the suggestion by Blackett and others that the relationship between the magnetic moment and the angular momentum of a rotating body represents 'some new and fundamental property of rotating matter'. In Chapman's 1948 review of solar magnetism he does not mention Elsasser or even dynamo models as we know them at all, and the preface to his 1948 second edition with Julian Bartels of the book *Geomagnetism* says that no changes were made because nothing of any significance had occurred since the 1940 edition. Joseph Larmor had proposed in 1919 that sunspots are generated by a steady dynamo effect from the solar rotation operating in a feedback with any slightest existing magnetic field to generate a stronger field, but Thomas Cowling in 1933 showed that such an axisymmetric mechanism would not explain the observed behaviour of the Sun. That was where things stood until Elsasser's works in 1946–7.

12 In philosophy this stance is called objectivity, a belief in mind-independent truths. Most scientists adhere to such a belief, but the existence of truth 'out there' is denied by many philosophers.

13 Newton's own assessment of his work was apparently more modest: 'I seem to have been only like a boy playing on the sea-shore, and diverting myself in now and then finding a smoother pebble or a prettier shell than ordinary, whilst the great ocean of truth lay all undiscovered before me.'

14 Nowadays, the *Kepler* spacecraft is famous for having discovered thousands of planets around other stars, with thousands of additional *Kepler* Objects of Interest (KOIs), probable planets but with a need to check for false positives. As of this writing, the main mission of the *Kepler* spacecraft is over, since it lost full three-axis gyroscopic control, but an extended K2 mission is continuing its quest to discover planets around other stars.

15 Experiments with people wearing inverting glasses show that after a few days of seeing the world inverted the brain adjusts and they see the world right-side up. When the glasses are removed, they see an inverted world for a while but their vision quickly returns to normal.

16 For scientific purposes, we often use 'gratings', pieces of glass or plastic with lines ruled very closely together, hundreds of thousands of parallel lines within each inch. Gratings also disperse light into its component colours. This happens because the light hitting the surface reflects from each of the grooves and combines either constructively or destructively, depending on the ratio of the line spacing in the grating to the wavelength of the light, peaking at a particular angle. For a given line spacing in the grating, the different wavelengths of light come out at different angles.

17 These boundaries are commonly used, but they are arbitrary. The spectrum of wavelengths actually has no limits at either end, other than practical restrictions involving the conditions, such as the power of the transmitter or the size of the receiver's antenna, under which the waves are produced and detected.

18 The pointing control system still being used for solar sounding rockets is known as SPARCS. This originally stood for Solar Pointing Aerobee Rocket Control System. Eventually, as other types of rocket were brought on line, the 'A' was changed to Attitude, keeping the acronym the same but broadening the meaning.

19 The U.S. military also launched solar satellites for research (and intelligence-gathering), especially the series of U.S. Navy SOLRAD satellites.

20 The visible band corresponds roughly to the wavelengths where the bulk of the Sun's energy is emitted and it is interesting to speculate about why this correspondence exists. Certainly it is not a coincidence and it seems likely that microbes, plants and animals evolved to take advantage of the available light. But that in itself does not explain why our atmosphere just happens to be transparent at those wavelengths. It may be the case that the living organisms on the Earth have also influenced the relative abundances of the constituent atoms and molecules of the atmosphere to help make it better suited to their existence.

21 There are now numerous space agencies launching satellites, the largest being in the U.S., Europe (ESA), the Russian Federation (Roscosmos) and Japan (JAXA). An up-to-date list of NASA satellites, including their status ('development', 'operating' or 'past') can be found at http://science.nasa.gov/heliophysics/missions. An interactive display of all orbiting satellites can be found at http://qz.com/296941/interactive-graphic-every-active-satellite-orbiting-earth.

22 It is easy, in retrospect, to accuse people on the wrong side of an argument of having been needlessly obtuse. Keep in mind though that while you are trying to figure out the correct answer, you do not yet know who is right, and healthy debate is a good thing. Those who raise objections usually have good reasons for doing so, and they provide stability by preventing every new (and usually incorrect) idea from being immediately taken as truth. The question of when one admits that the new view is correct and gives up an old theory is complex and has engendered many long discussions among philosophers.

23 These two types of observation are called, respectively, 'remote sensing' and 'in situ measurement'.

24 Its companion, *Voyager 2*, was launched sixteen days earlier, on 20 August, on a trajectory that took longer to reach Jupiter and Saturn but allowed it to encounter two other planets, Uranus and Neptune.

FURTHER READING

The following is a selection of both non-technical books and articles, and some technical articles for the reader who would like to pursue in more detail the subjects we discuss. We list first some recent books that are of general interest and then books that are more specifically relevant to the subjects treated in each of our chapters.

GENERAL INTEREST SOLAR BOOKS

Alexander, David, *The Sun* (Santa Barbara, CA, 2009). One of the Greenwood Guides to the Universe.

Berman, Bob, *The Sun's Heartbeat: And Other Stories from the Life of the Star that Powers Our Planet* (New York, 2011)

Bhatnagar, Arvind, and William C. Livingston, *Fundamentals of Solar Astronomy* (Singapore, 2005). Comprehensive and phenomenological but relatively non-mathematical.

Carlowicz, Michael J., and Ramon E. Lopez, *Storms from the Sun: The Emerging Science of Space Weather* (Washington, DC, 2000)

Golub, Leon, and Jay M. Pasachoff, *Nearest Star: The Surprising Science of Our Sun*, 2nd edn (New York, 2014). A non-technical trade book.

Haigh, Joanna D., and Peter Cargill, *The Sun's Influence on Climate* (Princeton, NJ, 2015)

Jago, Lucy, *Northern Lights* (New York, 2001)

Lang, Kenneth R., *Sun, Earth and Sky*, 2nd edn [2006], paperback reprint (New York, 2014)

—, *The Sun from Space* (New York, 2009)

Meadows, A. J., *Early Solar Physics* (London, 1970)

Mulvihill, Mary, *Lab Coats and Lace: The Lives and Legacies of Inspiring Irish Women Scientists and Pioneers* (Dublin, 2009)

Pasachoff, Jay M., *The Complete Idiot's Guide to the Sun* (Indianapolis, IN, 2003). Downloadable.

Zirker, Jack B., *Journey from the Center of the Sun* (Princeton, NJ, 2001; paperback, 2004)
—, *The Magnetic Universe: The Elusive Traces of an Invisible Force* (Baltimore, MD, 2009)
—, *Sunquakes: Probing the Interior of the Sun* (Princeton, NJ, 2003)

1 SUNSPOTS

Choudhuri, Arnab Rai, *Nature's Third Cycle: A Story of Sunspots* (Oxford, 2015)
Hale, George Ellery, *The New Heavens* [1922] (Charleston, NC, 2015)
Olson, Roberta J. M., and Jay M. Pasachoff, 'The Comets of Caroline Herschel (1750–1848), Sleuth of the Skies at Slough', *The Inspiration of Astronomical Phenomena* VII (insap.org) (Bath, 2010); published in *Culture and Cosmos*, XVI/1–2 (2012), pp. 53–76, also at http://arxiv.org/abs/1212.0809
Zhentao, Xu, 'Solar Observations in Ancient China and Solar Variability', *Philosophical Transactions of the Royal Society of London*, Series A, Mathematical and Physical Sciences, CCCXXX/1615 (1990) p. 513, DOI: 10.1098/rsta.1990.0032

2 LOOKING INSIDE THE SUN

Malin, S.R.C., and E. Bullard, 'The Direction of the Earth's Magnetic Field at London, 1570–1975', *Philosophical Transactions of the Royal Society of London*, Series A, CCLXXXXIX (1981), p. 357
Oldham's report on the 1897 earthquake is available for download at: https://archive.org/details/reportongreatea000ldhgoog
Ulrich, Roger, 'The Five-minute Oscillations on the Solar Surface', *Astrophysics Journal*, CLXII/3 (1970), p. 993

3 A SOLAR PULSE

King, Henry C., *History of the Telescope* (New York, 2011)
Memoirs of the Royal Astronomical Society, vol. XXVI (1856–7), p. 197
Sabra, A. I., *Theories of Light from Descartes to Newton* (Cambridge, 1981)

4 A SPECTRUM AND WHAT IT TELLS US

Comte, Auguste, *The Positive Philosophy* [1842], Book II, Chapter 1
Pasachoff, Jay M., 'The Hertzsprung–Russell Diagram', in *Discoveries in Modern Science: Exploration, Invention, Technology*, ed. James Trefil (Farmington Hills, MI, 2015), pp. 474–8

Pasachoff, Jay M., 'The H–R Diagram's 100th Anniversary', *Sky & Telescope* (June 2014), pp. 32–7

Smith, A. Mark, *From Sight to Light* (Chicago, IL, 2014)

5 THE SOLAR CHROMOSPHERE AND PROMINENCES

Foukal, Peter, and Jack Eddy, 'Did the Sun's Prairie Ever Stop Burning?' *Solar Physics*, CCXLV/2(2007), pp. 247–9, DOI: 10.1007/S11207-007-9057-8

6 THE VISIBLE CORONA

Baron, David, *American Eclipse: Thomas Edison and the Celestial Event that Illuminated a Nation* (New York, 2017)

Espenak, Fred, *Thousand Year Canon of Solar Eclipses: 1501 to 2500*, (Astropixels, 2014), www.astropixels.com. Maps and tables.

Espenak, Pat and Fred, 'Get Eclipsed': The Complete Guide to the American Eclipse (incl. a pair of partial-eclipse glasses), $6.00, http://astropixels.com/pubs/GetEclipsed.html

Guillermier, Pierre, and Serge Koutchmy, *Total Eclipses: Science, Observations, Myths and Legends* (New York, 1999)

Kepler, Johannes, *De Stella nova in pede Serpentarii (On the New Star in the Ophiuchus's Foot)* (Prague, 1606)

Littmann, Mark, and Fred Espenak, *Totality: The Great American Eclipses of 2017 and 2024* (Oxford, forthcoming 2017)

Littmann, Mark, Fred Espenak and Ken Willcox, *Totality: Eclipses of the Sun*, 3rd edn (Oxford, 2009)

Nath, Biman B., *The Discovery of Helium and the Birth of Astrophysics* (Charleston, NC, 2012)

Nordgren, Tyler, *Sun Moon Earth* (New York, 2016), http://tylernordgren.com

Peter, Hardi, and Bhola N. Dwivedi, 'Discovery of the Sun's Million-degree Hot Corona', *Astronomy and Space Sciences* (30 July 2014), http://dx.doi.org/10.3389/fspas.2014.00002

Zeiler, Michael, 'See the Great American Eclipse of August 21, 2017' (incl. 2 partial-eclipse glasses), http://greatamericaneclipse.com

7 THE INVISIBLE CORONA: A DISCUSSION MOSTLY ABOUT PHOTONS

Golub, Leon, and Jay M. Pasachoff, *The Solar Corona*, 2nd edn (Cambridge, 2010)

Mandel'štam, S. L., 'X-ray Emission of the Sun', *Space Science Reviews*, IV/5–6 (1965), p. 587.

8 STORMS FROM THE SUN: A DISCUSSION MOSTLY ABOUT PARTICLES AND FIELDS

Clark, Stuart, *The Sun Kings* (Princeton, NJ, 2007)

Cliver, E. W., 'Solar Activity and Geomagnetic Storms: The Corpuscular Hypothesis', *EOS: Transactions of the American Geophysical Union*, LXXV/609 (1994b)

—, 'Solar Activity and Geomagnetic Storms: The First 40 Years', *EOS: Transactions of the American Geophysical Union*, LXXV/569 (1994a)

—, 'Solar Activity and Geomagnetic Storms: From M Regions and Flares to Coronal Holes and CMEs', *EOS*, LXXVI/8 (1995), pp. 75–83

—, 'Was the Eclipse Comet of 1893 a Disconnected Coronal Mass Ejection?', *Solar Physics*, CXXII/2 (1989), p. 319

Crooker, N. U., and E. W. Cliver, 'Postmodern View of M-regions', *Journal of Geophysical Research*, XCIX (1994), p. 23383

Wulf, Andrea, *The Invention of Nature: Alexander von Humboldt's New World* (New York, 2015)

APPENDIX I: OBSERVING THE SUN SAFELY

Chou, B. Ralph, in Fred Espenak and Jay Anderson, *Eclipse Bulletin: Total Solar Eclipse of 2017 August 21* (Astropixels, 2015), www.astropixels.com, pp. 99–103

Pasachoff, Jay M., *A Field Guide to the Stars and Planets*, 4th edn, The Peterson Field Guide Series (Boston, MA, 2000; updated in 2016)

Pasachoff, Jay M., and Michael Covington, *The Cambridge Eclipse Photography Guide* (Cambridge, 1993)

APPENDIX II: AMATEUR SOLAR OBSERVING

Russo, Kate, *Total Addiction: The Life of an Eclipse Chaser* [ebook] (Charleston, NC, 2012)

—, *Totality: The Total Solar Eclipse of 2012 in Far North Queensland*, fcproductions. com.au, published by the author (2013)

APPENDIX III: OBSERVING THE CORONA AT ECLIPSE

See references for Chapter Six and Appendix I

Appendix IV: Observing the Sun From Space

Bester, A., *The Life and Death of a Satellite: A Biography of the Men and Machines at War with Peace* (Boston, MA, 1966)

Acknowledgements

LG thanks the people at NASA Headquarters, Goddard Space Flight Center and Marshall Space Flight Center who have supported the work described in this book; the management of the Smithsonian Astrophysical Observatory and its High Energy Astrophysics Division; and the many colleagues throughout the United States and worldwide who have made so many contributions to solar physics. I thank the people who read through portions or all of this manuscript and provided numerous comments and corrections: Anne Davenport, Jessica Law and Jenna Samra for helpful comments; and William Hanna, Michael Leaman and especially Peter Morris for their careful reading of the entire manuscript. For their support and encouragement I am grateful most especially to Anne; and to Jessica, Casey, Ansel and Ada; Pablo and Liz; Charles, Jessica and Jacob; and Manuel, Karla and Carlos.

JMP memorializes the late Donald H. Menzel, then Director of the Harvard College Observatory, for introducing him to solar eclipses when he was a freshman, 65 eclipses ago. We later worked together at the 1970 eclipse when I was a Menzel Research Fellow in the Harvard College Observatory. My work at the Big Bear Solar Observatory, then of the California Institute of Technology, with Harold Zirin furthered my solar education. Williams College has supported my activities, including my responsibility to attend and to study solar eclipses, for my decades there. My eclipse and other solar research has been supported over the years by the Committee for Research and Exploration of the National Geographic Society with a series of research grants, by the National Science Foundation (most recently by its Atmospheric and Geospace Sciences Division) and a series of expedition and research grants from the National Aeronautics and Space Administration. I thank Caltech, its Planetary Sciences Department and Prof. Andrew Ingersoll for sabbatical hospitality. I thank Naomi Pasachoff, Deborah Pasachoff/Ian Kezsbom and Eloise Pasachoff/Tom Glaisyer for editing and family support and for eclipse participation, for Deborah starting at age six months and for Eloise at age two and a half.

Photo Acknowledgements

The authors and publishers wish to express their thanks to the below sources of illustrative material and/or permission to reproduce it. A list of the acronyms employed can be found at the foot.

American Museum of Natural History Library, New York: 60; from H. W. Babcock, 1961, *The Astrophysical Journal*, CXXXIII/572 (© AAS, reproduced with permission): 24; © Wendy Carlos 2001 from individual images © 2001 by Jay M. Pasachoff: 61; © 2008 Miloslav Druckmüller, Peter Aniol, Martin Dietzel and Vojtech Rušin: 51; ESA/NASA/SOHO with the LASCO, NRL: 58; ESA/NASA/SWOOPS: 76; from Joseph Fraunhofer, *Bestimmung des Brechungs- und Farbenzerstreuungs-Vermögens verschiedener Glasarten* (Munich, [1817]): 35; from Galileo Galilei, *Istoria e Dimostrazioni Intorno Alle Macchie Solari e Loro Accidenti* [History and Demonstrations Concerning Sunspots and their Properties] (Rome, 1613): 3; from William Gilbert, *De Magnete, Magnetisque Cororibus et Magno Magnete* (London, 1600): 7; L. Golub drawings/diagrams: 8, 11, 12, 22, 23; L. Golub and NASA/GSFC/SDO: 69; L. Golub (SAO), Eberhard Spiller (IBM), and NASA: 65; GONG/NSA/AURA/NSF: 13, 15; GSFC/SDO/AIA, SAO and LMSAL: 62; from G. E. Hale, 1908, *The Astrophysical Journal*, XXVIII/100 (© AAS, reproduced with permission): 5; from G. E. Hale, 1919, *The Astrophysical Journal*, XLIX/153 (© AAS, reproduced with permission): 6 (top): from Edmond Halley, *A Description of the Passage of the Shadow of the Moon, over England, in the Total Eclipse of the Sun, on the 22d. Day of April 1715 in the Morning* (London, 1715): 55; from Edmond Halley, *Tabulæ Astronomicæ, Accedunt de Usu Tabularum Præcepta* (London, 1749): 19; courtesy David Hathaway, NASA/ARC: 20, 26; courtesy of Frank Hill, the NOAO and the NSO/GONG: 10; Hubble Heritage Team (AURA/STSCI, C. R. O'Dell (Vanderbilt), NASA /ESA: 85; JAXA/NASA/ESA Hinode/EIS: 70; JAXA/NASA/ESA/SAO: 17; JAXA/NASA/Hinode SOT, LMSAL: 47, 48; Ruth Kneale/NSO/NSF: 50; Serge Koutchmy, Institut d'Astrophysique, Paris/CNRS: 49; Serge Koutchmy and E. Tavabi, Institut d'Astrophysique, Paris and Sorbonne University: 46; Françoise Launay, Institute d'Astrophysique, Paris/CNRS: 43; Marshall Space Flight Center (NASA): 66; image processing by Christoforos Mouraditis: 42; Museum of Jurassic Technology, Culver City, California: 44, 59; N. A. Sharp (now NSF) National Solar Observatory; NOAO/NSO/Kitt Peak FTS/AURA/NSF: 36, 37; NASA: 67, 98; NASA/ESA/SOHO/MDI: 14; NASA/ESA/SOHO/MRL: 78, 79; NASA/GSFC/ Magnetospheric Multiscale (MMS) Mission: 80; NASA/GSFC/MAVEN: 82; NASA/ GSFC SDO: 97; NASA/GSFC/SDO/AIA: 72, 96; NASA/JPL-Caltech/GSFC/JAXA: 71; NASA/MSFC/STEREO: 83; NASA/SDO/AIA: 73, 77; NASA/SDO, AIA and HMI: 68; NASA/SDO/AIA/LMSAL: 75; NASA SDO/AIA and SDO/HMI/Stanford-Lockheed Institute for Space Research: 74; NASA/SDO/Stanford Lockheed Institute for Space Research/HMI: 21; NASA/VOYAGER, JPL-Caltech: 84; Natural Resources of Canada, Geological Survey of Canada: 18; National Center for Atmospheric Research/High Altitude Observatory (NCAR/HAO)/Mauna Loa Solar Observatory, courtesy of Joan Burkepile: 57; from Isaac Newton, *Opticks: or, A Treatise of the Reflexions, Refractions, Inflexions and Colours of Light* (London, 1704): 29; NSO/AURA/NSF: 5 (bottom); Deborah D. Pasachoff: 86, 88a, 88b, 89; photos Jay M. Pasachoff: 28, 32, 33, 52; photo Jay M. Pasachoff (courtesy of Wellesley College): 30; Jay M. Pasachoff, Allen B. Davis, and Vojtech Rušin, with processing by Miloslav Druckmüller: 39; from Jay M. Pasachoff and Alex Filippenko, *The Cosmos: Astronomy in the New Millennium*, 4th edn (Cambridge, 2014): 15, 38; Jay M. Pasachoff, composite by Leon Golub: 94; Jay M. Pasachoff, composite by Muzhou Lu, http://totalsolareclipse.org: 95; collection of Jay and Naomi Pasachoff: 3, 19, 31, 35, 40, 41; Jay and Naomi Pasachoff Collection, on deposit at The Chapin Library, Williams College; images courtesy of Wayne Hammond: 7, 29, 54, 55; Jay M. Pasachoff, Glenn Schneider, Dale Gary, Bin Chen, and Claude Plymate at the Big Bear Solar Observatory, New Jersey Institute of Technology: 1; Jay M. Pasachoff, Williams College Eclipse Expedition; funded in part by the Committee for Research and Exploration of the National Geographic Society, http://totalsolareclipse.org: 93; from [Jacob Pflaum], *Usum huius opusculi breviter exponemus. In principio offert se calendarium in 12 menses partitum* (Ulm, 1478): 54; Stephen W. Ramsden, Charlie Bates Solar Astronomy Project: 45; from P. A. Secchi, *Die Sonne: die wichtigen neuen Entdeckungen über ihren Bau, ihre Strahlungen, ihre Stellung im Weltall und ihr Verhaltniss zu den ubrigen Himmelskorpern* (Braunschweig, 1872): 40; Ralph Smith, Cairns, Australia: 4; SOHO (EIT), ESA/NASA: 27; courtesy of the SOHO/MDI consortium: 9; Space Weather Prediction Center, NOAA/NWS: 81; Swedish Solar Telescope (Swedish Research Council): 2; Aris Voulgaris of the Aristotle University of Thessaloniki, Greece, as part of the Williams College Eclipse Expedition, supported by a grant from the Committee for Research & Exploration of the National Geographic Society: 42, 56; Wikipedia/Mika Hirai, Williams College, modified from Bhamer, public domain: 63; Williams College-Hopkins Observatory http://astronomy.williams.edu/hopkins-observatory/eclipse-viewers/, from the collection of Luke Cole, donated by Nancy Shelby: 87, 91, 92; from William Hyde Wollaston, *Philosophical Transactions of the Royal Society of London*, LXXXXII (January 1802) [Fig. 3/Plate XIV, p. 380]: 34; from C. A. Young, *The Sun* (New York, 1881): 41.

Acronyms

AAS = American Astronomical Society

AIA = Atmospheric Imaging Assembly

ARC = Ames Research Center (NASA)

AURA = Association of Universities for Research in Astronomy

CNRS = Centre National de la Recherche Scientifique

DOOFAAS: Dumb Or Overly Forced Astronomical Acronyms Site
(www.cfa.harvard.edu/~gpetitpas/Links/Astroacro.html)

EIS = Extreme-ultraviolet Imaging Spectrometer

EIT = Extreme ultraviolet Imaging Telescope

ESA = European Space Agency

FTS = Fourier Transform Spectrometer

GONG = Global Oscillation Network Group

GSFC = Goddard Space Flight Center (NASA)

HAO = High Altitude Observatory (NCAR)

HMI = Helioseismic and Magnetic Imager

JAXA = Japan Aerospace Exploration Agency

JPL = Jet Propulsion Laboratory (NASA)

LASCO = Large And Spectrometric COronagraph Experiment

LM = Lockeed Martin Advanced Technology Center

LMSAL = Lockheed Martin Solar & Astrophysics Laboratory

MAVEN = Mars Atmosphere and Volatile EvolutioN mission

MDI = Michelson Doppler Imager (MDI was a project of the Stanford-Lockheed
Institute for Space Research and was a joint effort of the SOI in the W. W. Hansen
Experimental Physics Laboratory of Stanford University and the Solar and
Astrophysics Laboratory of the Lockheed-Martin Advanced Technology Center)

MELCO = Mitsubishi Electric Corporation

MMS = Magnetospheric MultiScale

MSFC = Marshall Space Flight Center (NASA)

NAOJ = National Astronomical Observatory of Japan

NASA = National Aeronautics and Space Administration

NCAR = National Center for Atmospheric Research

NOAA = National Oceanic and Atmospheric Administration

NOAO = National Optical Astronomy Observatory

NRL = Naval Research Laboratory

NSF = National Science Federation

NSO = National Solar Observatory

NWS = National Weather Service

OTA = Optical Telescope Assembly (the OTA was built, tested, and calibrated at the
Advanced Technology Center of the NAOJ with MELCO)

SAO = Smithsonian Astrophysical Observatory

SDO = Solar Dynamics Observatory

SOHO = Solar and Heliospheric Observatory (SOHO is a project of international cooperation between the ESA and the NASA with MDI)

SOI = Solar Oscillations Investigation

SOT = Solar Optical Telescope (the SOT is designed and developed by the international collaboration between NAOJ, LM, MELCO, the HAO (NCAR), MSFC (NASA), and JAXA)

SST = Swedish 1-m Solar Telescope (SST is now operated by the Institute for Solar Physics, Stockholm University, under the Swedish Research Council)

STEREO = Solar TErrestrial RElations Observatory

STS = Space Telescope Science Institute

SWOOPS = Solar Wind Observations Over the Poles of the Sun

INDEX

illustration numbers are in **bold italic**